Agricultural Engineering: Principles and Applications

Agricultural Engineering: Principles and Applications

Edited by
Cade Doherty

Larsen & Keller
www.larsen-keller.com

Agricultural Engineering: Principles and Applications
Edited by Cade Doherty
ISBN: 978-1-63549-017-6 (Hardback)

◲ Larsen & Keller

Published by Larsen and Keller Education,
5 Penn Plaza,
19th Floor,
New York, NY 10001, USA

Cataloging-in-Publication Data

Agricultural engineering: principles and applications / edited by Cade Doherty.
    p. cm.
Includes bibliographical references and index.
ISBN 978-1-63549-017-6
1. Agricultural engineering. 2. Agriculture.
I. Doherty, Cade.

S675 .A37 2017
631.3--dc23

The publisher's policy is to use permanent paper from mills that operate a sustainable forestry policy. Furthermore, the publisher ensures that the text paper and cover boards used have met acceptable environmental accreditation standards.

Printed and bound in the United States of America.

For more information regarding Larsen and Keller Education and its products, please visit the publisher's website www.larsen-keller.com

Table of Contents

Preface

The aim of this text is to present the topic of agricultural engineering in the most comprehensive way. It talks about the different methods and concepts in this area. Agricultural engineering deals with the design and creation of technologies and systems that regulate as well as improve existing utilization of resources and irrigation practices. The topics covered in this textbook offer the readers new insights into the field of agricultural engineering. It picks up individual branches and explains their need and contribution in the context of the growth of this field. Coherent flow of topics, student-friendly language and extensive use of examples make this book an invaluable source of knowledge.

A detailed account of the significant topics covered in this book is provided below:

Chapter 1- The cultivation of animals, medicinal plants as well as common plants and fungi for food is referred to as agriculture. Agricultural engineers are particularly concerned with the mechanization of agricultural machinery and make efficient resource management process. This chapter will provide an integrated understanding of agriculture and agricultural engineering.

Chapter 2- Intensive farming is a type of agriculture that has higher levels of input and output as per the agricultural land area. The methods and techniques used in intensive farming are managed intensive rotational grazing, crop rotation, irrigation, weed control and aquaculture. The topics discussed in the chapter are of great importance to broaden the existing knowledge on intensive farming.

Chapter 3- Agricultural machinery is the machinery that is used in either farming or any other agricultural practice. Some of the types of machines explained in this section are manure spreaders, cultivators, ploughs, drip irrigation, hog oiler and bulk tanks. The topics discussed in the section are of great importance to broaden the existing knowledge on agricultural machinery.

Chapter 4- Agricultural robots are used in agriculture; they mainly serve in the harvesting stage of the process. These agricultural robots are designed specifically for replacing human labor. The kinds of robots used are fruit picking robots, driverless sprayers and sheep shearing robots. This chapter will provide an integrated understanding on agricultural robots.

Chapter 5- Monoculture is the practice of growing a particular crop at a time. It is criticized by many for harming the environment and for putting the food supply chain at risk. The other various terminologies used in agricultural engineering are dosing, fertigation, convertible husbandry, monocropping, windrow and foodshed. The chapter serves as a source to understand the major terms used in agricultural engineering.

Chapter 6- The allied fields related to agricultural engineering are agricultural chemistry, agricultural diversification, agricultural economics, agricultural philosophy, agroecology, agrophysics etc. Agricultural chemistry is the study of chemistry and biochemistry; they are both equally important for the production of agriculture. This section discusses the allied fields of agricultural engineering in a critical manner providing key analysis to the subject matter.

I would like to make a special mention of my publisher who considered me worthy of this opportunity and also supported me throughout the process. I would also like to thank the editing team at the back-end who extended their help whenever required.

Editor

Introduction to Agriculture and Agricultural Engineering

The cultivation of animals, medicinal plants as well as common plants and fungi for food is referred to as agriculture. Agricultural engineers are particularly concerned with the mechanization of agricultural machinery and make efficient resource management process. This chapter will provide an integrated understanding of agriculture and agricultural engineering.

Agriculture

Fields in Záhorie (Slovakia) – a typical Central European agricultural region

Agriculture is the cultivation of animals, plants and fungi for food, fiber, biofuel, medicinal plants and other products used to sustain and enhance human life. Agriculture was the key development in the rise of sedentary human civilization, whereby farming of domesticated species created food surpluses that nurtured the development of civilization. The study of agriculture is known as agricultural science. The history of agriculture dates back thousands of years, and its development has been driven and defined by greatly different climates, cultures, and technologies. Industrial agriculture based on large-scale monoculture farming has become the dominant agricultural methodology.

Modern agronomy, plant breeding, agrochemicals such as pesticides and fertilizers, and technological developments have in many cases sharply increased yields from cultivation, but at the same time have caused widespread ecological damage and negative human health effects. Selective breeding and modern practices in animal husbandry have similarly increased the output of meat, but have raised concerns about animal welfare and the health effects of the antibiotics, growth hormones, and other chemicals commonly used in industrial meat production. Genetically modified organisms are an increasing component of agriculture, although they are banned in several countries. Agricultural food production and water management are increasingly becoming global issues that are fostering debate on a number of fronts. Significant degradation of land and water resources, including the depletion of aquifers, has been observed in recent decades, and the effects of global warming on agriculture and of agriculture on global warming are still not fully understood.

Domestic sheep and a cow (heifer) pastured together in South Africa

The major agricultural products can be broadly grouped into foods, fibers, fuels, and raw materials. Specific foods include cereals (grains), vegetables, fruits, oils, meats and spices. Fibers include cotton, wool, hemp, silk and flax. Raw materials include lumber and bamboo. Other useful materials are also produced by plants, such as resins, dyes, drugs, perfumes, biofuels and ornamental products such as cut flowers and nursery plants. Over one third of the world's workers are employed in agriculture, second only to the service sector, although the percentages of agricultural workers in developed countries has decreased significantly over the past several centuries.

Etymology and Terminology

The word *agriculture* is a late Middle English adaptation of Latin *agricultūra*, from *ager*, "field", and *cultūra*, "cultivation" or "growing". Agriculture usually refers to human activities, although it is also observed in certain species of ant, termite and ambrosia beetle. To practice agriculture means to use natural resources to "produce commodities which maintain life, including food, fiber, forest products, horticultural crops, and their related services." This definition includes arable farming or agronomy, and horticulture, all terms for the growing of plants, animal husbandry and forestry. A distinction is sometimes made between forestry and agriculture, based on the former's longer management rotations, extensive versus intensive management practices and development mainly by nature, rather than by man. Even then, it is acknowledged that there is a large amount of knowledge transfer and overlap between silviculture (the management of forests) and agriculture. In traditional farming, the two are often combined even on small landholdings, leading to the term agroforestry.

History

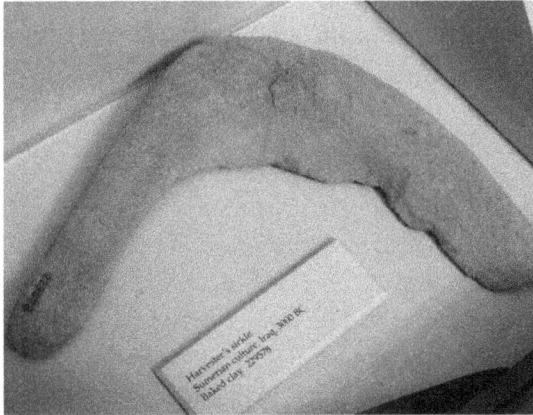

A Sumerian harvester's sickle made from baked clay (c. 3000 BC)

Agriculture began independently in different parts of the globe, and included a diverse range of taxa. At least 11 separate regions of the Old and New World were involved as independent centers of origin. Wild grains were collected and eaten from at least 105,000 years ago. Pigs were domesticated in Mesopotamia around 15,000 years ago. Rice was domesticated in China between 13,500 and 8,200 years ago, followed by mung, soy and azuki beans. Sheep were domesticated in Mesopotamia between 13,000 and 11,000 years ago. From around 11,500 years ago, the eight Neolithic founder crops, emmer and einkorn wheat, hulled barley, peas, lentils, bitter vetch, chick peas and flax were cultivated in the Levant. Cattle were domesticated from the wild aurochs in the areas of modern Turkey and Pakistan some 10,500 years ago. In the Andes of South America, the potato was domesticated between 10,000 and 7,000 years ago, along with beans, coca, llamas, alpacas, and guinea pigs. Sugarcane and some root vegetables were domesticated in New Guinea around 9,000 years ago. Sorghum was domesticated in the Sahel region of Africa by 7,000 years ago. Cotton was domesticated in Peru by 5,600 years ago, and was independently domesticated in Eurasia at an unknown time. In Mesoamerica, wild teosinte was domesticated to maize by 6,000 years ago.

In the Middle Ages, both in the Islamic world and in Europe, agriculture was transformed with improved techniques and the diffusion of crop plants, including the introduction of sugar, rice, cotton and fruit trees such as the orange to Europe by way of Al-Andalus. After 1492, the Columbian exchange brought New World crops such as maize, potatoes, sweet potatoes and manioc to Europe, and Old World crops such as wheat, barley, rice and turnips, and livestock including horses, cattle, sheep and goats to the Americas. Irrigation, crop rotation, and fertilizers were introduced soon after the Neolithic Revolution and developed much further in the past 200 years, starting with the British Agricultural Revolution. Since 1900, agriculture in the developed nations, and to a lesser extent in the developing world, has seen large rises in productivity as human labor has been replaced by mechanization, and assisted by synthetic fertilizers, pesticides, and selective breeding. The Haber-Bosch method allowed the synthesis of ammonium nitrate fertilizer on an industrial scale, greatly increasing crop yields. Modern agriculture has raised political issues including water pollution, biofuels, genetically modified organisms, tariffs and farm subsidies, leading to alternative approaches such as the organic movement.

Agriculture and Civilization

Civilization was the product of the Agricultural Neolithic Revolution; as H. G. Wells put it, "civilization was the agricultural surplus." In the course of history, civilization coincided in space with fertile areas such as The Fertile Crescent, and states formed mainly in circumscribed agricultural lands. The Great Wall of China and the Roman empire's *limes* (borders) demarcated the same northern frontier of cereal agriculture. This cereal belt fed the civilizations formed in the Axial Age and connected by the Silk Road.

Ancient Egyptians, whose agriculture depended exclusively on the Nile, deified the river, worshipped, and exalted it in a great hymn. The Chinese imperial court issued numerous edicts, stating: "Agriculture is the foundation of this Empire." Egyptian, Mesopotamian, Chinese, and Inca Emperors themselves plowed ceremonial fields in order to show personal example to everyone.

Ancient strategists, Chinese Guan Zhong and Shang Yang and Indian Kautilya, drew doctrines linking agriculture with military power. Agriculture defined the limits on how large and for how long an army could be mobilized. Shang Yang called agriculture and war the *One*. In the vast human pantheon of agricultural deities there are several deities who combined the functions of agriculture and war.

As the Neolithic Agricultural Revolution produced civilization, the modern Agricultural Revolution, begun in Britain (British Agricultural Revolution), made possible the Industrial civilization. The first precondition for industry was greater yields by less manpower, resulting in greater percentage of manpower available for non-agricultural sectors.

Types of Agriculture

Reindeer herds form the basis of pastoral agriculture for several Arctic and Subarctic peoples.

Pastoralism involves managing domesticated animals. In nomadic pastoralism, herds of livestock are moved from place to place in search of pasture, fodder, and water. This type of farming is practised in arid and semi-arid regions of Sahara, Central Asia and some parts of India.

In shifting cultivation, a small area of a forest is cleared by cutting down all the trees and the area is burned. The land is then used for growing crops for several years. When the soil becomes less fertile, the area is then abandoned. Another patch of land is selected and the process is repeated. This type of farming is practiced mainly in areas with abundant rainfall where the forest regenerates quickly. This practice is used in Northeast India, Southeast Asia, and the Amazon Basin.

Subsistence farming is practiced to satisfy family or local needs alone, with little left over for transport elsewhere. It is intensively practiced in Monsoon Asia and South-East Asia.

In intensive farming, the crops are cultivated for commercial purpose i.e., for selling. The main motive of the farmer is to make profit, with a low fallow ratio and a high use of inputs. This type of farming is mainly practiced in highly developed countries.

Contemporary Agriculture

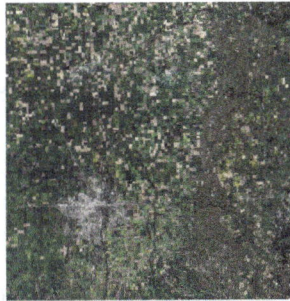

Satellite image of farming in Minnesota

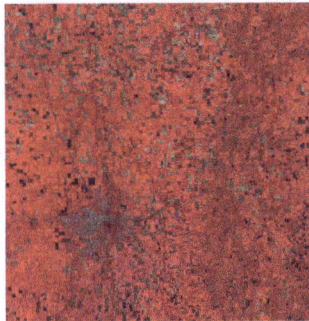

Infrared image of the above farms. Various colors indicate healthy crops (red), flooding (black) and unwanted pesticides (brown).

In the past century, agriculture has been characterized by increased productivity, the substitution of synthetic fertilizers and pesticides for labor, water pollution, and farm subsidies. In recent years there has been a backlash against the external environmental effects of conventional agriculture, resulting in the organic and sustainable agriculture movements. One of the major forces behind this movement has been the European Union, which first certified organic food in 1991 and began reform of its Common Agricultural Policy (CAP) in 2005 to phase out commodity-linked farm subsidies, also known as decoupling. The growth of organic farming has renewed research in alternative technologies such as integrated pest management and selective breeding. Recent mainstream technological developments include genetically modified food.

In 2007, higher incentives for farmers to grow non-food biofuel crops combined with other factors, such as over development of former farm lands, rising transportation costs, climate change, growing consumer demand in China and India, and population growth, caused food shortages in Asia, the Middle East, Africa, and Mexico, as well as rising food prices around the globe. As of December 2007, 37 countries faced food crises, and 20 had imposed some sort of food-price controls. Some of these shortages resulted in food riots and even deadly stampedes. The International Fund for Agricultural Development posits that an increase in smallholder agriculture may be part of the solution to concerns about food prices and overall food security. They in part base this on the experience of Vietnam, which went from a food importer to large food exporter and saw a significant drop in poverty, due mainly to the development of smallholder agriculture in the country.

Disease and land degradation are two of the major concerns in agriculture today. For example, an epidemic of stem rust on wheat caused by the Ug99 lineage is currently spreading across Africa and into Asia and is causing major concerns due to crop losses of 70% or more under some conditions. Approximately 40% of the world's agricultural land is seriously degraded. In Africa, if current trends of soil degradation continue, the continent might be able to feed just 25% of its population by 2025, according to United Nations University's Ghana-based Institute for Natural Resources in Africa.

Agrarian structure is a long-term structure in the Braudelian understanding of the concept. On a larger scale the agrarian structure is more dependent on the regional, social, cultural and historical factors than on the state's undertaken activities. Like in Poland, where despite running an intense agrarian policy for many years, the agrarian structure in 2002 has much in common with that found in 1921 soon after the partitions period.

In 2009, the agricultural output of China was the largest in the world, followed by the European Union, India and the United States, according to the International Monetary Fund. Economists measure the total factor productivity of agriculture and by this measure agriculture in the United States is roughly 1.7 times more productive than it was in 1948.

Workforce

As of 2011, the International Labour Organization states that approximately one billion people, or over 1/3 of the available work force, are employed in the global agricultural sector. Agriculture constitutes approximately 70% of the global employment of children, and in many countries employs the largest percentage of women of any industry. The service sector only overtook the agricultural sector as the largest global employer in 2007. Between 1997 and 2007, the percentage of people employed in agriculture fell by over four percentage points, a trend that is expected to continue. The number of people employed in agriculture varies widely on a per-country basis, ranging from less than 2% in countries like the US and Canada to over 80% in many African nations. In developed countries, these figures are significantly lower than in previous centuries. During the 16th century in Europe, for example, between 55 and 75 percent of the population was engaged in agriculture, depending on the country. By the 19th century in Europe, this had dropped to between 35 and 65 percent. In the same countries today, the figure is less than 10%.

Safety

Rollover protection bar on a Fordson tractor

Agriculture, specifically farming, remains a hazardous industry, and farmers worldwide remain at high risk of work-related injuries, lung disease, noise-induced hearing loss, skin diseases, as well as certain cancers related to chemical use and prolonged sun exposure. On industrialized farms, injuries frequently involve the use of agricultural machinery, and a common cause of fatal agricultural injuries in developed countries is tractor rollovers. Pesticides and other chemicals used in farming can also be hazardous to worker health, and workers exposed to pesticides may experience illness or have children with birth defects. As an industry in which families commonly share in work and live on the farm itself, entire families can be at risk for injuries, illness, and death. Common causes of fatal injuries among young farm workers include drowning, machinery and motor vehicle-related accidents.

The International Labour Organization considers agriculture "one of the most hazardous of all economic sectors." It estimates that the annual work-related death toll among agricultural employees is at least 170,000, twice the average rate of other jobs. In addition, incidences of death, injury and illness related to agricultural activities often go unreported. The organization has developed the Safety and Health in Agriculture

Convention, 2001, which covers the range of risks in the agriculture occupation, the prevention of these risks and the role that individuals and organizations engaged in agriculture should play.

Agricultural Production Systems

Crop Cultivation Systems

Rice cultivation in Bihar, India

Cropping systems vary among farms depending on the available resources and constraints; geography and climate of the farm; government policy; economic, social and political pressures; and the philosophy and culture of the farmer.

Shifting cultivation (or slash and burn) is a system in which forests are burnt, releasing nutrients to support cultivation of annual and then perennial crops for a period of several years. Then the plot is left fallow to regrow forest, and the farmer moves to a new plot, returning after many more years (10 – 20). This fallow period is shortened if population density grows, requiring the input of nutrients (fertilizer or manure) and some manual pest control. Annual cultivation is the next phase of intensity in which there is no fallow period. This requires even greater nutrient and pest control inputs.

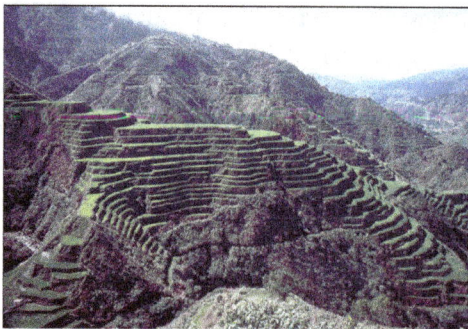

The Banaue Rice Terraces in Ifugao, Philippines

Further industrialization led to the use of monocultures, when one cultivar is planted on a large acreage. Because of the low biodiversity, nutrient use is uniform and pests tend to build up, necessitating the greater use of pesticides and fertilizers. Multiple cropping, in which several crops are grown sequentially in one year, and intercropping, when several crops are grown at the same time, are other kinds of annual cropping systems known as polycultures.

In subtropical and arid environments, the timing and extent of agriculture may be limited by rainfall, either not allowing multiple annual crops in a year, or requiring irrigation. In all of these environments perennial crops are grown (coffee, chocolate) and systems are practiced such as agroforestry. In temperate environments, where ecosystems were predominantly grassland or prairie, highly productive annual farming is the dominant agricultural system.

Crop Statistics

Important categories of crops include cereals and pseudocereals, pulses (legumes), forage, and fruits and vegetables. Specific crops are cultivated in distinct growing regions throughout the world. In millions of metric tons, based on FAO estimate.

Top agricultural products, by crop types (million tonnes) 2004 data	
Cereals	2,263
Vegetables and melons	866
Roots and tubers	715
Milk	619
Fruit	503
Meat	259
Oilcrops	133
Fish (2001 estimate)	130
Eggs	63
Pulses	60
Vegetable fiber	30
Source: Food and Agriculture Organization (FAO)	

Top agricultural products, by individual crops (million tonnes) 2011 data	
Sugar cane	1794
Maize	883
Rice	722
Wheat	704
Potatoes	374
Sugar beet	271
Soybeans	260
Cassava	252
Tomatoes	159
Barley	134
Source: Food and Agriculture Organization (FAO)	

Livestock Production Systems

Ploughing rice paddy fields with water buffalo, in Indonesia

Animals, including horses, mules, oxen, water buffalo, camels, llamas, alpacas, donkeys, and dogs, are often used to help cultivate fields, harvest crops, wrangle other animals, and transport farm products to buyers. Animal husbandry not only refers to the breeding and raising of animals for meat or to harvest animal products (like milk, eggs, or wool) on a continual basis, but also to the breeding and care of species for work and companionship.

An ox-pulled plough in India

Livestock production systems can be defined based on feed source, as grassland-based, mixed, and landless. As of 2010, 30% of Earth's ice- and water-free area was used for producing livestock, with the sector employing approximately 1.3 billion people. Between the 1960s and the 2000s, there was a significant increase in livestock production, both by numbers and by carcass weight, especially among beef, pigs and chickens, the latter of which had production increased by almost a factor of 10. Non-meat animals, such as milk cows and egg-producing chickens, also showed significant production increases. Global cattle, sheep and goat populations are expected to continue to increase sharply through 2050. Aquaculture or fish farming, the production of fish for human consumption in confined operations, is one of the fastest growing sectors of food production, growing at an average of 9% a year between 1975 and 2007.

During the second half of the 20th century, producers using selective breeding focused on creating livestock breeds and crossbreeds that increased production, while mostly disregarding the need to preserve genetic diversity. This trend has led to a significant decrease in genetic diversity and resources among livestock breeds, leading to a corresponding decrease in disease resistance and local adaptations previously found among traditional breeds.

Grassland based livestock production relies upon plant material such as shrubland, rangeland, and pastures for feeding ruminant animals. Outside nutrient inputs may be used, however manure is returned directly to the grassland as a major nutrient source. This system is particularly important in areas where crop production is not feasible because of climate or soil, representing 30 – 40 million pastoralists. Mixed production systems use grassland, fodder crops and grain feed crops as feed for ruminant and monogastric (one stomach; mainly chickens and pigs) livestock. Manure is typically recycled in mixed systems as a fertilizer for crops.

Landless systems rely upon feed from outside the farm, representing the de-linking of crop and livestock production found more prevalently in Organisation for Economic Co-operation and Development(OECD) member countries. Synthetic fertilizers are more heavily relied upon for crop production and manure utilization becomes a challenge as well as a source for pollution. Industrialized countries use these operations to produce much of the global supplies of poultry and pork. Scientists estimate that 75% of the growth in livestock production between 2003 and 2030 will be in confined animal feeding operations, sometimes called factory farming. Much of this growth is happening in developing countries in Asia, with much smaller amounts of growth in Africa. Some of the practices used in commercial livestock production, including the usage of growth hormones, are controversial.

Production Practices

Road leading across the farm allows machinery access to the farm for production practices

Farming is the practice of agriculture by specialized labor in an area primarily devoted to agricultural processes, in service of a dislocated population usually in a city.

Tillage is the practice of plowing soil to prepare for planting or for nutrient incorporation or for pest control. Tillage varies in intensity from conventional to no-till. It may improve productivity by warming the soil, incorporating fertilizer and controlling weeds, but also renders soil more prone to erosion, triggers the decomposition of organic matter releasing CO_2, and reduces the abundance and diversity of soil organisms.

Pest control includes the management of weeds, insects, mites, and diseases. Chemical (pesticides), biological (biocontrol), mechanical (tillage), and cultural practices are used. Cultural practices include crop rotation, culling, cover crops, intercropping, composting, avoidance, and resistance. Integrated pest management attempts to use all of these methods to keep pest populations below the number which would cause economic loss, and recommends pesticides as a last resort.

Nutrient management includes both the source of nutrient inputs for crop and livestock production, and the method of utilization of manure produced by livestock. Nutrient inputs can be chemical inorganic fertilizers, manure, green manure, compost and mined minerals. Crop nutrient use may also be managed using cultural techniques such as crop rotation or a fallow period. Manure is used either by holding livestock where the feed crop is growing, such as in managed intensive rotational grazing, or by spreading either dry or liquid formulations of manure on cropland or pastures.

Water management is needed where rainfall is insufficient or variable, which occurs to some degree in most regions of the world. Some farmers use irrigation to supplement rainfall. In other areas such as the Great Plains in the U.S. and Canada, farmers use a fallow year to conserve soil moisture to use for growing a crop in the following year. Agriculture represents 70% of freshwater use worldwide.

According to a report by the International Food Policy Research Institute, agricultural technologies will have the greatest impact on food production if adopted in combination with each other; using a model that assessed how eleven technologies could impact agricultural productivity, food security and trade by 2050, the International Food Policy Research Institute found that the number of people at risk from hunger could be reduced by as much as 40% and food prices could be reduced by almost half.

"Payment for ecosystem services (PES) can further incentivise efforts to green the agriculture sector. This is an approach that verifies values and rewards the benefits of ecosystem services provided by green agricultural practices." "Innovative PES measures could include reforestation payments made by cities to upstream communities in rural areas of shared watersheds for improved quantities and quality of fresh water for municipal users. Ecoservice payments by farmers to upstream forest stewards for properly managing the flow of soil nutrients, and methods to monetise the carbon sequestration and emission reduction credit benefits of green agriculture practices in order to compensate farmers for their efforts to restore and build SOM and employ other practices."

Crop Alteration and Biotechnology

Crop alteration has been practiced by humankind for thousands of years, since the beginning of civilization. Altering crops through breeding practices changes the genetic make-up of a plant to develop crops with more beneficial characteristics for humans, for example, larger fruits or seeds, drought-tolerance, or resistance to pests. Significant advances in plant breeding ensued after the work of geneticist Gregor Mendel. His work on dominant and recessive alleles, although initially largely ignored for almost 50 years, gave plant breeders a better understanding of genetics and breeding techniques. Crop breeding includes techniques such as plant selection with desirable traits, self-pollination and cross-pollination, and molecular techniques that genetically modify the organism.

Tractor and chaser bin

Domestication of plants has, over the centuries increased yield, improved disease resistance and drought tolerance, eased harvest and improved the taste and nutritional value of crop plants. Careful selection and breeding have had enormous effects on the characteristics of crop plants. Plant selection and breeding in the 1920s and 1930s improved pasture (grasses and clover) in New Zealand. Extensive X-ray and ultraviolet induced mutagenesis efforts (i.e. primitive genetic engineering) during the 1950s produced the modern commercial varieties of grains such as wheat, corn (maize) and barley.

The Green Revolution popularized the use of conventional hybridization to sharply increase yield by creating "high-yielding varieties". For example, average yields of corn (maize) in the USA have increased from around 2.5 tons per hectare (t/ha) (40 bushels per acre) in 1900 to about 9.4 t/ha (150 bushels per acre) in 2001. Similarly, worldwide average wheat yields have increased from less than 1 t/ha in 1900 to more than 2.5 t/ha in 1990. South American average wheat yields are around 2 t/ha, African under 1 t/ha, and Egypt and Arabia up to 3.5 to 4 t/ha with irrigation. In contrast, the average wheat yield in countries such as France is over 8 t/ha. Variations in yields are due mainly to variation in climate, genetics, and the level of intensive farming techniques (use of fertilizers, chemical pest control, growth control to avoid lodging).

Genetic Engineering

Genetically modified organisms (GMO) are organisms whose genetic material has been altered by genetic engineering techniques generally known as recombinant DNA technology. Genetic engineering has expanded the genes available to breeders to utilize in creating desired germlines for new crops. Increased durability, nutritional content, insect and virus resistance and herbicide tolerance are a few of the attributes bred into crops through genetic engineering. For some, GMO crops cause food safety and food labeling concerns. Numerous countries have placed restrictions on the production, import or use of GMO foods and crops, which have been put in place due to concerns over potential health issues, declining agricultural diversity and contamination of non-GMO crops. Currently a global treaty, the Biosafety Protocol, regulates the trade of GMOs. There is ongoing discussion regarding the labeling of foods made from GMOs, and while the EU currently requires all GMO foods to be labeled, the US does not.

Herbicide-resistant seed has a gene implanted into its genome that allows the plants to tolerate exposure to herbicides, including glyphosates. These seeds allow the farmer to grow a crop that can be sprayed with herbicides to control weeds without harming the resistant crop. Herbicide-tolerant crops are used by farmers worldwide. With the increasing use of herbicide-tolerant crops, comes an increase in the use of glyphosate-based herbicide sprays. In some areas glyphosate resistant weeds have developed, causing farmers to switch to other herbicides. Some studies also link widespread glyphosate usage to iron deficiencies in some crops, which is both a crop production and a nutritional quality concern, with potential economic and health implications.

Other GMO crops used by growers include insect-resistant crops, which have a gene from the soil bacterium *Bacillus thuringiensis* (Bt), which produces a toxin specific to insects. These crops protect plants from damage by insects. Some believe that similar or better pest-resistance traits can be acquired through traditional breeding practices, and resistance to various pests can be gained through hybridization or cross-pollination with wild species. In some cases, wild species are the primary source of resistance traits; some tomato cultivars that have gained resistance to at least 19 diseases did so through crossing with wild populations of tomatoes.

Environmental Impact

Water pollution in a rural stream due to runoff from farming activity in New Zealand

Agriculture, as implemented through the method of farming, imposes external costs upon society through pesticides, nutrient runoff, excessive water usage, loss of natural environment and assorted other problems. A 2000 assessment of agriculture in the UK determined total external costs for 1996 of £2,343 million, or £208 per hectare. A 2005 analysis of these costs in the USA concluded that cropland imposes approximately $5 to 16 billion ($30 to $96 per hectare), while livestock production imposes $714 million. Both studies, which focused solely on the fiscal impacts, concluded that more should be done to internalize external costs. Neither included subsidies in their analysis, but they noted that subsidies also influence the cost of agriculture to society. In 2010, the International Resource Panel of the United Nations Environment Programme published a report assessing the environmental impacts of consumption and production. The study found that agriculture and food consumption are two of the most important drivers of environmental pressures, particularly habitat change, climate change, water use and toxic emissions. The 2011 UNEP Green Economy report states that "[a]gricultural operations, excluding land use changes, produce approximately 13 per cent of anthropogenic global GHG emissions. This includes GHGs emitted by the use of inorganic fertilisers agro-chemical pesticides and herbicides; (GHG emissions resulting from production of these inputs are included in industrial emissions); and fossil fuel-energy inputs. "On average we find that the total amount of fresh residues from agricultural and forestry production for second- generation biofuel production amounts to 3.8 billion tonnes per year between 2011 and 2050 (with an average annual growth rate of 11 per cent throughout the period analysed, accounting for higher growth during early years, 48 per cent for 2011–2020 and an average 2 per cent annual expansion after 2020)."

Livestock Issues

A senior UN official and co-author of a UN report detailing this problem, Henning Steinfeld, said "Livestock are one of the most significant contributors to today's most serious environmental problems". Livestock production occupies 70% of all land used for agriculture, or 30% of the land surface of the planet. It is one of the largest sources of greenhouse gases, responsible for 18% of the world's greenhouse gas emissions as measured in CO_2 equivalents. By comparison, all transportation emits 13.5% of the CO_2. It produces 65% of human-related nitrous oxide (which has 296 times the global warming potential of CO_2) and 37% of all human-induced methane (which is 23 times as warming as CO_2.) It also generates 64% of the ammonia emission. Livestock expansion is cited as a key factor driving deforestation; in the Amazon basin 70% of previously forested area is now occupied by pastures and the remainder used for feedcrops. Through deforestation and land degradation, livestock is also driving reductions in biodiversity. Furthermore, the UNEP states that "methane emissions from global livestock are projected to increase by 60 per cent by 2030 under current practices and consumption patterns."

Land and Water Issues

Land transformation, the use of land to yield goods and services, is the most substantial way humans alter the Earth's ecosystems, and is considered the driving force in the loss of biodiversity. Estimates of the amount of land transformed by humans vary from 39 to 50%. Land degradation, the long-term decline in ecosystem function and productivity, is estimated to be occurring on 24% of land worldwide, with cropland overrepresented. The UN-FAO report cites land management as the driving factor behind degradation and reports that 1.5 billion people rely upon the degrading land. Degradation can be deforestation, desertification, soil erosion, mineral depletion, or chemical degradation (acidification and salinization).

Eutrophication, excessive nutrients in aquatic ecosystems resulting in algal blooms and anoxia, leads to fish kills, loss of biodiversity, and renders water unfit for drinking and other industrial uses. Excessive fertilization and manure application to cropland, as well as high livestock stocking densities cause nutrient (mainly nitrogen and phosphorus) runoff and leaching from agricultural land. These nutrients are major nonpoint pollutants contributing to eutrophication of aquatic ecosystems.

Agriculture accounts for 70 percent of withdrawals of freshwater resources. Agriculture is a major draw on water from aquifers, and currently draws from those underground water sources at an unsustainable rate. It is long known that aquifers in areas as diverse as northern China, the Upper Ganges and the western US are being depleted, and new research extends these problems to aquifers in Iran, Mexico and Saudi Arabia. Increasing pressure is being placed on water resources by industry and urban areas, meaning that water scarcity is increasing and agriculture is facing the challenge of producing more food for the world's growing population with reduced water resources. Agricultural water usage can also cause major environmental problems, including the destruction of natural wetlands, the spread of water-borne diseases, and land degradation through salinization and waterlogging, when irrigation is performed incorrectly.

Pesticides

Pesticide use has increased since 1950 to 2.5 million short tons annually worldwide, yet crop loss from pests has remained relatively constant. The World Health Organization estimated in 1992 that 3 million pesticide poisonings occur annually, causing 220,000 deaths. Pesticides select for pesticide resistance in the pest population, leading to a condition termed the "pesticide treadmill" in which pest resistance warrants the development of a new pesticide.

An alternative argument is that the way to "save the environment" and prevent famine is by using pesticides and intensive high yield farming, a view exemplified by a quote heading the Center for Global Food Issues website: 'Growing more per acre leaves more land for nature'. However, critics argue that a trade-off between the environment and a need for food is not inevitable, and that pesticides simply replace good agronomic practices such as crop rotation. The UNEP introduces the Push–pull agricultural pest management technique which involves intercropping that uses plant aromas to repel or push away pests while pulling in or attracting the right insects. "The implementation of push-pull in eastern Africa has significantly increased maize yields and the combined cultivation of N-fixing forage crops has enriched the soil and has also provided farmers with feed for livestock. With increased livestock operations, the farmers are able to produce meat, milk and other dairy products and they use the manure as organic fertiliser that returns nutrients to the fields."

Climate Change

Climate change has the potential to affect agriculture through changes in temperature, rainfall (timing and quantity), CO_2, solar radiation and the interaction of these elements. Extreme events, such as droughts and floods, are forecast to increase as climate change takes hold. Agriculture is among sectors most vulnerable to the impacts of climate change; water supply for example, will be critical to sustain agricultural production and provide the increase in food output required to sustain the world's growing population. Fluctuations in the flow of rivers are likely to increase in the twenty-first century. Based on the experience of countries in the Nile river basin (Ethiopia, Kenya and Sudan) and other developing countries, depletion of water resources during seasons crucial for agriculture can lead to a decline in yield by up to 50%. Transformational approaches will be needed to manage natural resources in the future. For example, policies, practices and tools promoting climate-smart agriculture will be important, as will better use of scientific information on climate for assessing risks and vulnerability. Planners and policy-makers will need to help create suitable policies that encourage funding for such agricultural transformation.

Agriculture in its many forms can both mitigate or worsen global warming. Some of the increase in CO_2 in the atmosphere comes from the decomposition of organic matter in the soil, and much of the methane emitted into the atmosphere is caused by the decomposition of organic matter in wet soils such as rice paddy fields, as well as the normal digestive activities of farm animals. Further, wet or anaerobic soils also lose nitrogen through denitrification, releasing the greenhouse gases nitric oxide and nitrous oxide. Changes in management can reduce the release of these greenhouse gases, and soil can further be used to sequester some of the CO_2 in the atmosphere. Informed by the UNEP, "[a]griculture also produces about 58 per cent of global nitrous oxide emissions and about 47 per cent of global methane emissions. Cattle and rice farms release methane, fertilized fields release nitrous oxide, and the cutting down of rainforests to grow crops or raise livestock releases carbon dioxide. Both of these gases have a far greater global warming potential per tonne than CO2 (298 times and 25 times respectively)."

There are several factors within the field of agriculture that contribute to the large amount of CO2 emissions. The diversity of the sources ranges from the production of farming tools to the transport of harvested produce. Approximately 8% of the national carbon footprint is due to agricultural sources. Of that, 75% is of the carbon emissions released from the production of crop assisting chemicals. Factories producing insecticides, herbicides, fungicides, and fertilizers are a major culprit of the greenhouse gas. Productivity on the farm itself and the use of machinery is another source of the carbon emission. Almost all the industrial machines used in modern farming are powered by fossil fuels. These instruments are burning fossil fuels from the beginning of the process to the end. Tractors are the root of this source. The tractor is going to burn fuel and release CO_2 just to run. The amount of emissions from the machinery increase with the attachment of different units and need for more power. During the soil preparation stage tillers and plows will be used to disrupt the soil. During growth watering pumps and sprayers are used to keep the crops hydrated. And when the crops are ready for picking a forage or combine harvester is used. These types of machinery all require additional energy which leads to increased carbon dioxide emissions from the basic tractors. The final major contribution to CO_2 emissions in agriculture is in the final transport of produce. Local farming suffered a decline over the past century due to large amounts of farm subsidies. The majority of crops are shipped hundreds of miles to various processing plants before ending up in the grocery store. These shipments are made using fossil fuel burning modes of transportation. Inevitably these transport adds to carbon dioxide emissions.

Sustainability

Some major organizations are hailing farming within agroecosystems as the way forward for mainstream agriculture. Current farming methods have resulted in over-stretched water resources, high levels of erosion and reduced soil fertility. According to a report by the International Water Management Institute and UNEP, there is not enough water to continue farming using current practices; therefore how critical water, land, and ecosystem resources are used to boost crop yields must be reconsidered. The report suggested assigning value to ecosystems, recognizing environmental and livelihood tradeoffs, and balancing the rights of a variety of users and interests. Inequities that result when such measures are adopted would need to be addressed, such as the reallocation of water from poor to rich, the clearing of land to make way for more productive farmland, or the preservation of a wetland system that limits fishing rights.

Technological advancements help provide farmers with tools and resources to make farming more sustainable. New technologies have given rise to innovations like conservation tillage, a farming process which helps prevent land loss to erosion, water pollution and enhances carbon sequestration.

According to a report by the International Food Policy Research Institute (IFPRI), agricultural technologies will have the greatest impact on food production if adopted in combination with each other; using a model that assessed how eleven technologies could impact agricultural productivity, food security and trade by 2050, IFPRI found that the number of people at risk from hunger could be reduced by as much as 40% and food prices could be reduced by almost half.

Agricultural Economics

Agricultural economics refers to economics as it relates to the "production, distribution and consumption of [agricultural] goods and services". Combining agricultural production with general theories of marketing and business as a discipline of study began in the late 1800s, and grew significantly through the 20th century. Although the study of agricultural economics is relatively recent, major trends in agriculture have significantly affected national and international economies throughout history, ranging from tenant farmers and sharecropping in the post-American Civil War Southern United States to the European feudal system of manorialism. In the United States, and elsewhere, food costs attributed to food processing, distribution, and agricultural marketing, sometimes referred to as the value chain, have risen while the costs attributed to farming have declined. This is related to the greater efficiency of farming, combined with the increased level of value addition (e.g. more highly processed products) provided by the supply chain. Market concentration has increased in the sector as well, and although the total effect of the increased market concentration is likely increased efficiency, the changes redistribute economic surplus from producers (farmers) and consumers, and may have negative implications for rural communities.

National government policies can significantly change the economic marketplace for agricultural products, in the form of taxation, subsidies, tariffs and other measures. Since at least the 1960s, a combination of import/export restrictions, exchange rate policies and subsidies have affected farmers in both the developing and developed world. In the 1980s, it was clear that non-subsidized farmers in developing countries were experiencing adverse effects from national policies that created artificially low global prices for farm products. Between the mid-1980s and the early 2000s, several international agreements were put into place that limited agricultural tariffs, subsidies and other trade restrictions.

However, as of 2009, there was still a significant amount of policy-driven distortion in global agricultural product prices. The three agricultural products with the greatest amount of trade distortion were sugar, milk and rice, mainly due to taxation. Among the oilseeds, sesame had the greatest amount of taxation, but overall, feed grains and oilseeds had much lower levels of taxation than livestock products. Since the 1980s, policy-driven distortions have seen a greater decrease among livestock products than crops during the worldwide reforms in agricultural policy. Despite this progress, certain crops, such as cotton, still see subsidies in developed countries artificially deflating global prices, causing hardship in developing countries with non-subsidized farmers. Unprocessed commodities (i.e. corn, soybeans, cows) are generally graded to indicate quality. The quality affects the price the producer receives. Commodities are generally reported by production quantities, such as volume, number or weight.

Agricultural Science

Agricultural science is a broad multidisciplinary field of biology that encompasses the parts of exact, natural, economic and social sciences that are used in the practice and understanding of agriculture. (Veterinary science, but not animal science, is often excluded from the definition.)

Energy and Agriculture

Since the 1940s, agricultural productivity has increased dramatically, due largely to the increased use of energy-intensive mechanization, fertilizers and pesticides. The vast majority of this energy input comes from fossil fuel sources. Between the 1960–65 measuring cycle and the cycle from 1986 to 1990, the Green Revolution transformed agriculture around the globe, with world grain production increasing significantly (between 70% and 390% for wheat and 60% to 150% for rice, depending on geographic area) as world population doubled. Modern agriculture's heavy reliance on petrochemicals and mechanization has raised concerns that oil shortages could increase costs and reduce agricultural output, causing food shortages.

Agriculture and food system share (%) of total energy consumption by three industrialized nations			
Country	Year	Agriculture (direct & indirect)	Food system
United Kingdom	2005	1.9	11
United States	2002	2.0	14
Sweden	2000	2.5	13

Modern or industrialized agriculture is dependent on fossil fuels in two fundamental ways: 1. direct consumption on the farm and 2. indirect consumption to manufacture inputs used on the farm. Direct consumption includes the use of lubricants and fuels to operate farm vehicles and machinery; and use of gasoline, liquid propane, and electricity to power dryers, pumps, lights, heaters, and coolers. American farms directly consumed about 1.2 exajoules (1.1 quadrillion BTU) in 2002, or just over 1% of the nation's total energy.

Indirect consumption is mainly oil and natural gas used to manufacture fertilizers and pesticides, which accounted for 0.6 exajoules (0.6 quadrillion BTU) in 2002. The natural gas and coal consumed by the production of nitrogen fertilizer can account for over half of the agricultural energy usage. China utilizes mostly coal in the production of nitrogen fertilizer, while most of Europe uses large amounts of natural gas and small amounts of coal. According to a 2010 report published by The Royal Society, agriculture is increasingly dependent on the direct and indirect input of fossil fuels. Overall, the fuels used in agriculture vary based on several factors, including crop, production system and location. The energy used to manufacture farm machinery is also a form of indirect agricultural energy consumption. Together, direct and indirect consumption by US farms accounts for about 2% of the nation's energy use. Direct and indirect energy consumption by U.S. farms peaked in 1979, and has gradually declined over the past 30 years. Food systems encompass not just agricultural production, but also off-farm processing, packaging, transporting, marketing, consumption, and disposal of food and food-related items. Agriculture accounts for less than one-fifth of food system energy use in the US.

Mitigation of Effects of Petroleum Shortages

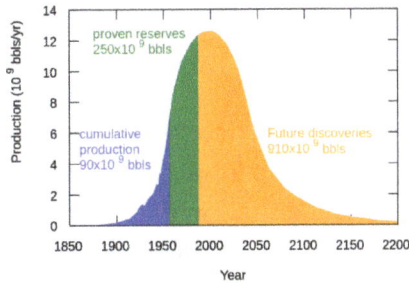

M. King Hubbert's prediction of world petroleum production rates. Modern agriculture is totally reliant on petroleum energy

In the event of a petroleum shortage, organic agriculture can be more attractive than conventional practices that use petroleum-based pesticides, herbicides, or fertilizers. Some studies using modern organic farming methods have reported yields equal to or higher than those available from conventional farming. In the aftermath of the fall of the Soviet Union, with shortages of conventional petroleum-based inputs, Cuba made use of mostly organic practices, including biopesticides, plant-based pesticides and sustainable cropping practices, to feed its populace. However, organic farming may be more labor-intensive and would require a shift of the workforce from urban to rural areas. The reconditioning of soil to restore organic matter lost during the use of monoculture agriculture techniques is important to provide a reservoir of plant-available nutrients, to maintain texture, and to minimize erosion.

It has been suggested that rural communities might obtain fuel from the biochar and synfuel process, which uses agricultural *waste* to provide charcoal fertilizer, some fuel *and* food, instead of the normal food vs. fuel debate. As the synfuel would be used on-site, the process would be more efficient and might just provide enough fuel for a new organic-agriculture fusion.

It has been suggested that some transgenic plants may some day be developed which would allow for maintaining or increasing yields while requiring fewer fossil-fuel-derived inputs than conventional crops. The possibility of success of these programs is questioned by ecologists and economists concerned with unsustainable GMO practices such as terminator seeds. While there has been some research on sustainability using GMO crops, at least one prominent multi-year attempt by Monsanto Company has been unsuccessful, though during the same period traditional breeding techniques yielded a more sustainable variety of the same crop.

Policy

Agricultural policy is the set of government decisions and actions relating to domestic agriculture and imports of foreign agricultural products. Governments usually implement agricultural policies with the goal of achieving a specific outcome in the domestic agricultural product markets. Some overarching themes include risk management and adjustment (including policies related to climate change, food safety and natural disasters), economic stability (including policies related to taxes), natural resources and environmental sustainability (especially water policy), research and development, and market access for domestic commodities (including relations with global organizations and agreements with other countries). Agricultural policy can also touch on food quality, ensuring that the food supply is of a consistent and known quality, food security, ensuring that the food supply meets the population's needs, and conservation. Policy programs can range from financial programs, such as subsidies, to encouraging producers to enroll in voluntary quality assurance programs.

United States farm subsidies in 2005

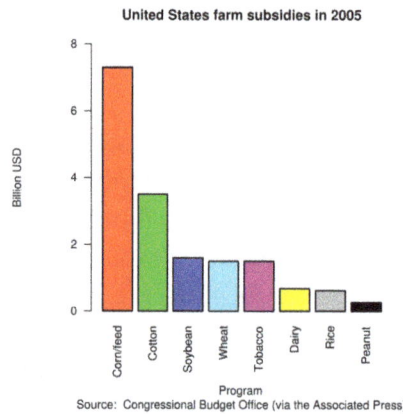

Source: Congressional Budget Office (via the Associated Press)

From a Congressional Budget Office report

There are many influences on the creation of agricultural policy, including consumers, agribusiness, trade lobbies and other groups. Agribusiness interests hold a large amount of influence over policy making, in the form of lobbying and campaign contributions. Political action groups, including those interested in environmental issues and labor unions, also provide influence, as do lobbying organizations representing individual agricultural commodities. The Food and Agriculture Organization of the United Nations (FAO) leads international efforts to defeat hunger and provides a forum for the negotiation of global agricultural regulations and agreements. Dr. Samuel Jutzi, director of FAO's animal production and health division, states that lobbying by large corporations has stopped reforms that would improve human health and the environment. For example, proposals in 2010 for a voluntary code of conduct for the livestock industry that would have provided incentives for improving standards for health, and environmental regulations, such as the number of animals an area of land can support without long-term damage, were successfully defeated due to large food company pressure.

Agricultural Engineering

Agricultural engineering in forestry

Specialties

Agricultural engineers may engage in any of the following areas:

- design of agricultural machinery, equipment, and agricultural structures
- internal combustion engines as applied to agricultural machinery
- agricultural resource management (including land use and water use)
- water management, conservation, and storage for crop irrigation and livestock production
- surveying and land profiling
- climatology and atmospheric science
- soil management and conservation, including erosion and erosion control
- seeding, tillage, harvesting, and processing of crops
- livestock production, including poultry, fish, and dairy animals
- waste management, including animal waste, agricultural residues, and fertilizer runoff
- food engineering and the processing of agricultural products
- basic principles of circuit analysis, as applied to electrical motors
- physical and chemical properties of materials used in, or produced by, agricultural production
- bioresource engineering, which uses machines on the molecular level to help the environment.
- Design of experiments related to crop and animal production

History

The first curriculum in agricultural engineering was established at Iowa State University by Professor J. B. Davidson in 1903. The American Society of Agricultural Engineers, now known as the American Society of Agricultural and Biological Engineers, was founded in 1907.

Agricultural Engineers

Agricultural engineers may perform tasks such as planning, supervising and managing the building of dairy effluent schemes, irrigation, drainage, flood water control systems, performing environmental impact assessments, agricultural product processing and interpret research results and implement relevant practices. A large percentage of agricultural engineers work in academia or for government agencies such as the United States Department of Agriculture or state agricultural extension services. Some are consultants, employed by private engineering firms, while others work in industry, for manufacturers of agricultural machinery, equipment, processing technology, and structures for housing livestock and storing crops. Agricultural engineers work in production, sales, management, research and development, or applied science.

In the United Kingdom the term Agricultural Engineer is often also used to describe a person that repairs or modifies agricultural equipment.

Academic Programs in Agricultural and Biosystems Engineering

Below is a listing of known academic programs that offer bachelor's degrees (B.S. or B.S.E. or B.E / B.Tech) in what ABET terms "Agricultural Engineering", "Biosystems Engineering", "Biological Engineering", or similarly named programs. ABET accredits college and university programs in the disciplines of applied science, computing, engineering, and engineering technology.

America

North America

Institution	Department	Web site	Email
Auburn University	Biosystems Engineering	www.eng.auburn.edu/	taylost at auburn.edu
Cal Poly San Luis Obispo	BioResource and Agricultural Engineering	www.brae.calpoly.edu/	ksolomon at calpoly.edu
Clemson University	Biosystems Engineering		jjacks4 at clemson.edu
Cornell University	Biological and Environmental Engineering	bee.cornell.edu/	baa7 at cornell.edu
Dalhousie University	Department of Engineering (Agricultural Campus)	www.dal.ca	Cathy.Wood at Dal.Ca
Iowa State University	Agricultural and Biosystems Engineering	www.abe.iastate.edu/	estaben at iastate.edu
Kansas State University	Biological and Agricultural Engineering	www.bae.ksu.edu/home	contact-l at bae.ksu.edu
Louisiana State University	Biological and Agricultural Engineering	www.bae.lsu.edu	DConstant at agcenter.lsu.edu
McGill University	Department of Bioresource Engineering	www.mcgill.ca/bioeng/	valerie.orsat at mcgill.ca

Michigan State University	Biosystems and Agricultural Engineering	www.egr.msu.edu/bae/	srivasta at egr.msu.edu
Mississippi State University	Agricultural and Biological Engineering	www.abe.msstate.edu	jpote at mafes.msstate.edu
North Carolina State University	Biological and Agricultural Engineering	www.bae.ncsu.edu/	robert_evans at ncsu.edu
North Carolina Agricultural & Technical State University	Chemical, Biological and Bioengineering Department	www.ncat.edu/	sbknisle at ncat.edu
North Dakota State University	Agricultural and Biosystems Engineering	www.ndsu.edu/aben/	aben at ndsu.edu
Ohio State University	Food, Agricultural and Biological Engineering	fabe.osu.edu	shearer.95 at osu.edu
Oklahoma State University	Biosystems and Agricultural Engineering	biosystems.okstate.edu	daniel.thomas at okstate.edu
Oregon State University	Biological and Ecological Engineering Department	bee.oregonstate.edu/	john.bolte at oregonstate.edu
Penn State University	Agricultural and Biosystems Engineering	abe.psu.edu/	hzh at psu.edu
Purdue University	Agricultural and Biological Engineering	www.purdue.edu/abe	engelb at purdue.edu
South Dakota State University	Agricultural and Biosystems Engineering	www.sdstate.edu/abe/	Van.Kelley at sdstate.edu
Texas A&M University	Biological and Agricultural Engineering	baen.tamu.edu/	info at baen.tamu.edu
University of Arizona	Agricultural and Biosystems Engineering	cals.arizona.edu/abe	slackd at u.arizona.edu
University of Arkansas	Biological and Agricultural Engineering	www.baeg.uark.edu/	lverma at uark.edu
University of California, Davis	Biological and Agricultural Engineering	bae.engineering.ucdavis.edu/	rhpiedrahita at ucdavis.edu
University of Florida	Agricultural and Biological Engineering	www.abe.ufl.edu/	dhaman at ufl.edu
University of Georgia	Biological and Agricultural Engineering	www.engr.uga.edu	donleo at engr.uga.edu
University of Illinois	Agricultural and Biological Engineering	abe.illinois.edu/	kcting at Illinois.edu
University of Kentucky	Biosystems and Agricultural Engineering	jokko.bae.uky.edu/BAEHome.asp	sue.nokes at uky.edu
University of Nebraska-Lincoln	Biological Systems Engineering	bse.unl.edu	mriley3 at unl.edu
University of Manitoba	Biosystems Engineering	http://umanitoba.ca/	headbio at cc.umanitoba.ca

University of Minnesota	Bioproducts and Bio-systems Engineering	www.bbe.umn.edu/	shri at umn.edu
University of Missouri	Biological Engineering	bioengineering.missouri.edu/	TanJ at missouri.edu
University of Saskatchewan	Biosystems Engineering and Soil Science	www.engr.usask.ca	Pat.Hunchak at usask.ca
University of Tennessee	Biosystems Engineering & Soil Science	bioengr.ag.utk.edu	edrumm at utk.edu
University of Wisconsin	Biological Systems Engineering	bse.wisc.edu/	doug.reinemann at wisc.edu
Utah State University	Biological Engineering	be.usu.edu	ron.sims at usu.edu
Virginia Polytechnic University	Biological Systems Engineering	www.bse.vt.edu	mlwolfe at vt.edu

Mexico, Central and South America

Institution	Department	Web site
Universidad Autónoma Chapingo, Mexico	Irrigation, Soils, Plant Science Department	portal.chapingo.mx/irrigacion/
Universidad Autónoma Agraria Antonio Narro, Mexico	Agricultural Engineering	http://www.uaaan.mx/v3/
Universidad de Costa Rica, Costa Rica	Agricultural and Biosystems Engineering	
National University of Colombia	Agricultural Engineering	www.ing.unal.edu.co/
University of Campinas, Brazil	Agricultural Engineering	www.unicamp.br/unicamp/
University of São Paulo, Brazil	Biosystems Engineering	
Federal University of Viçosa, Brazil	Agricultural Engineering	
Federal University of Lavras, Brazil	Engineering Department	
Federal University of Pelotas, Brazil	Engineering Centre	
Pontifical Catholic University of Chile	Faculty of Agronomy & Forest Engineering	http://agronomia.uc.cl/
University of Concepción (Campus Chillán), Chile	Agricultural Engineering	

Europe

Institution	Department	Web site
Agricultural Engineering	http://eng.au.dk/	
Agricultural University of Athens, Greece	Natural Resources Management and Agricultural Engineering	http://www.aua.gr/

Aristotle University of Thessaloniki, Greece	Hydraulics, Soil science and Agricultural Engineering	http://www.agro.auth.gr/
Ankara University, Turkey	Agricultural Engineering	agri.ankara.edu.tr
Akdeniz University, Turkey	Agricultural machinery and Technology Engineering	
Atatürk University, Turkey	Agricultural Structures and Irrigation	www.atauni.edu.tr
ETH Zurich, Switzerland	Agricultural Science	https://www.ethz.ch/
Harran University, Turkey	Agricultural Machinery	ziraat.harran.edu.tr
Harran University, Turkey	Agricultural Structures and Irrigation	ziraat.harran.edu.tr
Namik Kemal University, Turkey	Agricultural Engineering	http://ziraat-en.nku.edu.tr/
University of Bologna, Italy	Agricultural engineering and mechanics	http://www.unibo.it/
Universitat Politècnica de Catalunya, Spain	Agricultural Engineering	http://www.upc.edu/
Leibniz University Hannover, Germany	Water Resources and Environmental Management	http://www.bgt-hannover.de/
University of Hohenheim, Germany	Institute of Agricultural Engineering	www.uni-hohenheim.de/
University of Reading, UK	Agricultural Engineering	http://www.reading.ac.uk/
University of Liège, Belgium	Gembloux Agro-Bio Tech,	
Universiteit Gent, Belgium	Faculty of Bioscience Engineering	
K.U. Leuven, Belgium	Faculty of Bioscience Engineering	www.kuleuven.be/english
U.C. Louvain, Belgium	Faculty of Bioscience Engineering	
University of Lisbon, Portugal	Instituto Superior de Agronomia	www.isa.utl.pt/ University of Évora, Portugal
University of Algarve, Portugal	Faculty of Sciences and Technology	www.ualg.pt/home/en/curso/1459
University of Trás-os-Montes and Alto Douro, Portugal	Agronomy	http://ecav.utad.pt/vPT/Area2/Departamentos/agronomia

University of Natural Resources and Life Sciences, Austria	Sustainable Agricultural Systems	http://www.boku.ac.at/
University College Dublin, Ireland	Biosystems Engineering	www.ucd.ie/eacollege/biosystems/
University of Debrecen, Hungary	Agricultural Engineering	http://edu.unideb.hu/

Asia

Institution	Department	Web site
CCS Haryana Agricultural University,Hisar,India	Agricultural Engineering	http://hau.ernet.in/coaet/coaet.php
Nims University,Jaipur,India	Agricultural Engineering	http://nimsuniversity.org/schools/ nims-engineering-technology-schools/ nims-school-of-agricultural-engineering/
IIT KHARAGPUR	Agricultural Engineering	http://www.agri.iitkgp.ernet.in/
North Eastern Regional Institute of Science and Technology, India	Agricultural Engineering	http://www.nerist.ac.in
Acharya N G Ranga Agricultural University, Bapatla, India	Agricultural Engineering	www.caebapatla.co.in
Punjab Agricultural University,Ludhiana,Punjab,India	Agricultural Engineering	
G.B. Pant University of Agriculture & Technology, Pantnagar	Agricultural Engineering	
Nagaland University	Agricultural Engineering and Technology	
China Agricultural University	Agricultural Engineering	english.cau.edu.cn/col/col5470/index.html
Northwest Agriculture and Forestry University, China	Agricultural Soil and Water Engineering	en.nwsuaf.edu.cn/
Shanghai Jiatong University, China	Biological Engineering; Food Science and Engineering	en.sjtu.edu.cn/academics/undergraduate-programs/
Xi'an Jiaotong University, China	Energy and Power Engineering	nd.xjtu.edu.cn/web/English.htm
Yunnan Agricultural University, China	Agricultural Water-Soil Engineering	www.at0086.com/YUNNAU
Zhejiang University, China	Biosystems Engineering and Food Science	www.caefs.zju.edu.cn/en/index.asp

Odisha University of Agriculture & Technology	Agricultural Engineering	http://www.ouat.ac.in/
University of Agriculture Faisalabad Pakistan	Agricultural Engineering, Water Resources Engineering, Food Engineering, and Energy Engineering	http://www.uaf.edu.pk
Universiti Putra Malaysia, Malaysia	Biological and Agricultural Engineering	
Bogor Agricultural University, Indonesia	Mechanical and Biosystem Engineering, Food Science and Technology, Agroindustrial Technology, Civil and Environmental Engineering	
Universitas Gadjah Mada, Indonesia	Agricultural and Biosystems Engineering	
University of Southeastern Philippines	Agricultural Engineering	
Bangladesh Agricultural University	Agricultural Engineering and Food Engineering	
Sylhet Agricultural University	Agricultural Engineering and Technology	
University of Agricultural Sciences, Bengaluru,Karnataka, India	Agricultural Engineering	[www.uasbangalore.edu.in]
University of agricultural sciences, Raichur, Karnataka, India		
Southern Philippines Agri-Business and Marine and Aquatic School of Technology, Digos City, Davao del Sur	Agricultural Engineering	
Xavier University - Ateneo de Cagayan, Cagayan de Oro City, Philippines	Agricultural Engineering	
Khyber Pakhtunkhwa University of Engineering and Technology, Peshawar Pakistan	Agricultural Engineering, Soil and Water Conservation Engineering, Farm Machinery	

King Mongkut's Institute of Technology Ladkrabang (KMITL), Bangkok, Thailand	Agricultural Engineering	
Central Mindanao University - Maramag, Bukidnon	Agricultural Engineering	
A. S. ENGINEERING ENTERPRISE - RAMRAJATALA,	HOWRAH Agricultural Equipment	
Central Luzon State University (CLSU)	Agricultural and Biosystems Engineering	http://www.cenclsu.ph/
PJTS Agricultural University, Hyderabad	Agricultural Engineering	http://www.pjtsau.ac.in

|Bahawal-din-Zakriya University Multan,Pakistan(BZU) || Agricultural Engineering ||

Oceania

Institution	Department	Web site
University of Southern Queensland, Australia	School of Civil Engineering and Surveying, Faculty of Health, Engineering and Sciences	www.usq.edu.au
Flinders University, Australia	School of Computer Science, Engineering and Mathematics	flinders.edu.au

References

- Berg, Paul; Singer, Maxine (15 August 2003). George Beadle: An Uncommon Farmer. The Emergence of Genetics in the 20th century. Cold Springs Harbor Laboratory Press. ISBN 978-0-87969-688-7.

- Safety and health in agriculture. International Labour Organization. 1999. pp. 77–. ISBN 978-92-2-111517-5. Retrieved 13 September 2010.

- Chantrell, Glynnis, ed. (2002). The Oxford Dictionary of Word Histories. Oxford University Press. p. 14. ISBN 0-19-863121-9.

- Committee on Forestry Research, National Research Council (1990). Forestry Research: A Mandate for Change. National Academies Press. pp. 15–16. ISBN 0-309-04248-8.

- Budowski, Gerardo (1982). "Applicability of agro-forestry systems". In MacDonald, L.H. Agro-forestry in the African Humid Tropics. United Nations University. ISBN 92-808-0364-6. Retrieved 17 March 2016.

- Ensminger, M.E.; Parker, R.O. (1986). Sheep and Goat Science (Fifth ed.). Interstate Printers and Publishers. ISBN 0-8134-2464-X.

- Broudy, Eric (1979). The Book of Looms: A History of the Handloom from Ancient Times to the Present. UPNE. p. 81. ISBN 978-0-87451-649-4.

- National Geographic (2015). Food Journeys of a Lifetime. National Geographic Society. p. 126. ISBN 978-1-4262-1609-1.

- Harmon, Katherine (17 December 2009). "Humans feasting on grains for at least 100,000 years". Scientific American. Retrieved 28 August 2016.

- "Africa may be able to feed only 25% of its population by 2025". Mongabay. 14 December 2006. Archived from the original on 27 November 2011. Retrieved 15 July 2016.

- Stokstad, Marilyn (2005). Medieval Castles. Greenwood Publishing Group. ISBN 0-313-32525-1. Retrieved 17 March 2016.

- "World oil supplies are set to run out faster than expected, warn scientists". The Independent. 14 June 2007. Archived from the original on 21 October 2010. Retrieved 14 July 2016.

- Martin Heller; Gregory Keoleian (2000). "Life Cycle-Based Sustainability Indicators for Assessment of the U.S. Food System" (PDF). University of Michigan Center for Sustainable Food Systems. Retrieved 17 March 2016.

Intensive Farming: Methods and Techniques

Intensive farming is a type of agriculture that has higher levels of input and output as per the agricultural land area. The methods and techniques used in intensive farming are managed intensive rotational grazing, crop rotation, irrigation, weed control and aquaculture. The topics discussed in the chapter are of great importance to broaden the existing knowledge on intensive farming.

Intensive Farming

Intensive farming or intensive agriculture also known as industrial agriculture is characterized by a low fallow ratio and higher use of inputs such as capital and labour per unit land area. This is in contrast to traditional agriculture in which the inputs per unit land are lower.

Intensive animal husbandry involves either large numbers of animals raised on limited land, usually confined animal feeding operations (CAFO) often referred to as factory farms, or managed intensive rotational grazing (MIRG). Both increase the yields of food and fiber per acre as compared to traditional animal husbandry. In a CAFO feed is brought to the animals, which are seldom moved, while in MIRG the animals are repeatedly moved to fresh forage.

Intensive crop agriculture is characterised by innovations designed to increase yield. Techniques include planting multiple crops per year, reducing the frequency of fallow years and improving cultivars. It also involves increased use of fertilizers, plant growth regulators, pesticides and mechanization, controlled by increased and more detailed analysis of growing conditions, including weather, soil, water, weeds and pests.

This system is supported by ongoing innovation in agricultural machinery and farming methods, genetic technology, techniques for achieving economies of scale, logistics and data collection and analysis technology. Intensive farms are widespread in developed nations and increasingly prevalent worldwide. Most of the meat, dairy, eggs, fruits and vegetables available in supermarkets are produced by such farms.

Smaller intensive farms usually include higher inputs of labor and more often use sustainable intensive methods. The farming practices commonly found on such farms are referred to as appropriate technology. These farms are less widespread in both developed countries and worldwide, but are growing more rapidly. Most of the food available in specialty markets such as farmers markets is produced by these smallholder farms.

History

Early 20th-century image of a tractor ploughing an alfalfa field

Agricultural development in Britain between the 16th century and the mid-19th century saw a massive increase in agricultural productivity and net output. This in turn supported unprecedented population growth, freeing up a significant percentage of the workforce, and thereby helped enable the Industrial Revolution. Historians cited enclosure, mechanization, four-field crop rotation, and selective breeding as the most important innovations.

Industrial agriculture arose along with the Industrial Revolution. By the early 19th century, agricultural techniques, implements, seed stocks and cultivars had so improved that yield per land unit was many times that seen in the Middle Ages.

The industrialization phase involved a continuing process of mechanization. Horse-drawn machinery such as the McCormick reaper revolutionized harvesting, while inventions such as the cotton gin reduced the cost of processing. During this same period, farmers began to use steam-powered threshers and tractors, although they were expensive and dangerous.In 1892, the first gasoline-powered tractor was successfully developed, and in 1923, the International Harvester Farmall tractor became the first all-purpose tractor, marking an inflection point in the replacement of draft animals with machines. Mechanical harvesters (combines), planters, transplanters and other equipment were then developed, further revolutionizing agriculture. These inventions increased yields and allowed individual farmers to manage increasingly large farms.

The identification of nitrogen, potassium, and phosphorus (NPK) as critical factors in plant growth led to the manufacture of synthetic fertilizers, further increasing crop yields. In 1909 the Haber-Bosch method to synthesize ammonium nitrate was first demonstrated. NPK fertilizers stimulated the first concerns about industrial agriculture, due to concerns that they came with serious side effects such as soil compaction, soil erosion and declines in overall soil fertility, along with health concerns about toxic chemicals entering the food supply.

The identification of carbon as a critical factor in plant growth and soil health, particularly in the form of humus, led to so-called *sustainable agriculture*, alternative forms of intensive agriculture that also surpass traditional agriculture, without side effects or health issues. Farmers adopting this approach were initially referred to as *humus farmers*, later as *organic farmers*.

The discovery of vitamins and their role in nutrition, in the first two decades of the 20th century, led to vitamin supplements, which in the 1920s allowed some livestock to be raised indoors, reducing their exposure to adverse natural elements.Chemicals developed for use in World War II gave rise to synthetic pesticides.

Following World War II, synthetic fertilizer use increased rapidly, while sustainable intensive farming advanced much more slowly. Most of the resources in developed nations went to improving industrial intensive farming, and very little went to improving organic farming. Thus, particularly in the developed nations, industrial intensive farming grew to become the dominant form of agriculture.

The discovery of antibiotics and vaccines facilitated raising livestock in CAFOs by reducing diseases caused by crowding. Developments in logistics and refrigeration as well as processing technology made long-distance distribution feasible.

Between 1700 and 1980, "the total area of cultivated land worldwide increased 466%" and yields increased dramatically, particularly because of selectively bred high-yielding varieties, fertilizers, pesticides, irrigation and machinery. Global agricultural production doubled between 1820 1920; between 1920 and 1950; between 1950 and 1965; and again between 1965 and 1975 to feed a global population that grew from one billion in 1800 to 6.5 billion in 2002. The number of people involved in farming in industrial countries dropped, from 24 percent of the American population to 1.5 percent in 2002. In 1940, each farmworker supplied 11 consumers, whereas in 2002, each worker supplied 90 consumers. The number of farms also decreased and their ownership became more concentrated. In 2000 in the U.S., four companies produced 81 percent of cows, 73 percent of sheep, 57 percent of pigs, and 50 percent of chickens, cited as an example of "vertical integration" by the president of the U.S. National Farmers' Union. Between 1967 and 2002 the one million pig farms in America consolidated into 114,000 with 80 million pigs (out of 95 million) produced each year on factory farms, according to the U.S. National Pork Producers Council. According to the Worldwatch Institute, 74 percent of the world's poultry, 43 percent of beef, and 68 percent of eggs are produced this way.

Concerns over the sustainability of industrial agriculture, which has become associated with decreased soil quality, and over the environmental effects of fertilizers and pesticides, have not subsided. Alternatives such as integrated pest management (IPM) have had little impact because policies encourage the use of pesticides and IPM is knowledge-intensive. These concerns sustained the organic movement and caused a resurgence in sustainable intensive farming and funding for the development of appropriate technology.

Famines continued throughout the 20th century. Through the effects of climactic events, government policy, war and crop failure, millions of people died in each of at least ten famines between the 1920s and the 1990s.

Techniques and Technologies

Livestock

A commercial chicken house raising broiler pullets for meat.

Confined Animal Feeding Operations

Intensive livestock farming, also called "factory farming" is a term referring to the process of raising livestock in confinement at high stocking density. "Concentrated animal feeding operations" (CAFO) or "intensive livestock operations", can hold large numbers (some up to hundreds of thousands) of cows, hogs, turkeys or chickens, often indoors. The essence of such farms is the concentration of livestock in a given space. The aim is to provide maximum output at the lowest possible cost and with the greatest level of food safety. The term is often used pejoratively. However, CAFOs have dramatically increased the production of food from animal husbandry worldwide, both in terms of total food produced and efficiency.

Food and water is delivered to the animals, and therapeutic use of antimicrobial agents, vitamin supplements and growth hormones are often employed. Growth hormones are not used on chickens nor on any animal in the European Union. Undesirable behaviours often related to the stress of confinement led to a search for docile breeds (e.g., with natural dominance behaviours bred out), physical restraints to stop interaction, such as individual cages for chickens, or physically modification such as the de-beaking of chickens to reduce the harm of fighting.

The CAFO designation resulted from the 1972 US Federal Clean Water Act, which was enacted to protect and restore lakes and rivers to a "fishable, swimmable" quality. The

United States Environmental Protection Agency (EPA) identified certain animal feeding operations, along with many other types of industry, as "point source" groundwater polluters. These operations were subjected to regulation.

In 17 states in the U.S., isolated cases of groundwater contamination were linked to CAFOs. For example, the ten million hogs in North Carolina generate 19 million tons of waste per year. The U.S. federal government acknowledges the waste disposal issue and requires that animal waste be stored in lagoons. These lagoons can be as large as 7.5 acres (30,000 m²). Lagoons not protected with an impermeable liner can leak into groundwater under some conditions, as can runoff from manure used as fertilizer. A lagoon that burst in 1995 released 25 million gallons of nitrous sludge in North Carolina's New River. The spill allegedly killed eight to ten million fish.

The large concentration of animals, animal waste and dead animals in a small space poses ethical issues to some consumers. Animal rights and animal welfare activists have charged that intensive animal rearing is cruel to animals.

Other concerns include persistent noxious odor, the effects on human health and the role of antibiotics use in the rise of resistant infectious bacteria.

According to the U.S. Centers for Disease Control and Prevention (CDC), farms on which animals are intensively reared can cause adverse health reactions in farm workers. Workers may develop acute and/or chronic lung disease, musculoskeletal injuries and may catch (zoonotic) infections from the animals.

Managed Intensive Rotational Grazing

Managed Intensive Rotational Grazing (MIRG), also known as cell grazing, mob grazing and holistic managed planned grazing, is a variety of forage use in which herds/ flocks are regularly and systematically moved to fresh, rested grazing areas to maximize the quality and quantity of forage growth. MIRG can be used with cattle, sheep, goats, pigs, chickens, turkeys, ducks and other animals. The herds graze one portion of pasture, or a paddock, while allowing the others to recover. Resting grazed lands allows the vegetation to renew energy reserves, rebuild shoot systems, and deepen root systems, resulting in long-term maximum biomass production. MIRG is especially effective because grazers thrive on the more tender younger plant stems. MIRG also leave parasites behind to die off minimizing or eliminating the need for de-wormers. Pasture systems alone can allow grazers to meet their energy requirements, and with the increased productivity of MIRG systems, the animals obtain the majority of their nutritional needs, in some cases all, without the supplemental feed sources that are required in continuous grazing systems or CAFOs.

Crops

The Green Revolution transformed farming in many developing countries. It spread technologies that had already existed, but had not been widely used outside of industrialized nations. These technologies included "miracle seeds", pesticides, irrigation and synthetic nitrogen fertilizer.

Seeds

In the 1970s scientists created strains of maize, wheat, and rice that are generally referred to as high-yielding varieties (HYV). HYVs have an increased nitrogen-absorbing potential compared to other varieties. Since cereals that absorbed extra nitrogen would typically lodge (fall over) before harvest, semi-dwarfing genes were bred into their genomes. Norin 10 wheat, a variety developed by Orville Vogel from Japanese dwarf wheat varieties, was instrumental in developing wheat cultivars. IR8, the first widely implemented HYV rice to be developed by the International Rice Research Institute, was created through a cross between an Indonesian variety named "Peta" and a Chinese variety named "Dee Geo Woo Gen."

With the availability of molecular genetics in Arabidopsis and rice the mutant genes responsible (*reduced height (rht), gibberellin insensitive (gai1)* and *slender rice (slr1)*) have been cloned and identified as cellular signalling components of gibberellic acid, a phytohormone involved in regulating stem growth via its effect on cell division. Photosynthetic investment in the stem is reduced dramatically as the shorter plants are inherently more mechanically stable. Nutrients become redirected to grain production, amplifying in particular the yield effect of chemical fertilisers.

HYVs significantly outperform traditional varieties in the presence of adequate irrigation, pesticides and fertilizers. In the absence of these inputs, traditional varieties may outperform HYVs. They were developed as F1 hybrids, meaning seeds need to be purchased every season to obtain maximum benefit, thus increasing costs.

Crop Rotation

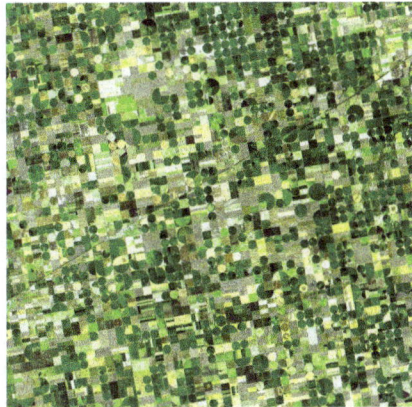

Satellite image of circular crop fields in Haskell County, Kansas in late June 2001. Healthy, growing crops of corn and sorghum are green (Sorghum may be slightly paler). Wheat is brilliant gold. Fields of brown have been recently harvested and plowed under or have lain in fallow for the year.

Crop rotation or crop sequencing is the practice of growing a series of dissimilar types of crops in the same space in sequential seasons for benefits such as avoiding pathogen and pest buildup that occurs when one species is continuously cropped. Crop rotation also seeks to balance the nutrient demands of various crops to avoid soil nutrient depletion. A traditional component of crop rotation is the replenishment of nitrogen through the use of legumes and green manure in sequence with cereals and other crops. Crop rotation can also improve soil structure and fertility by alternating deep-rooted and shallow-rooted plants. One technique is to plant multi-species cover crops between commercial crops. This combines the advantages of intensive farming with continuous cover and polyculture.

Irrigation

Overhead irrigation, center pivot designed

Crop irrigation accounts for 70% of the world's fresh water use.

Flood irrigation, the oldest and most common type, is typically unevenly distributed, as parts of a field may receive excess water in order to deliver sufficient quantities to other parts. Overhead irrigation, using center-pivot or lateral-moving sprinklers, gives a much more equal and controlled distribution pattern. Drip irrigation is the most expensive and least-used type, but delivers water to plant roots with minimal losses.

Water catchment management measures include recharge pits, which capture rainwater and runoff and use it to recharge groundwater supplies. This helps in the replenishment of groundwater wells and eventually reduces soil erosion. Dammed rivers creating Reservoirs store water for irrigation and other uses over large areas. Smaller areas sometimes use irrigation ponds or groundwater.

Weed Control

In agriculture, systematic weed management is usually required, often performed by machines such as cultivators or liquid herbicide sprayers. Herbicides kill specific targets while leaving the crop relatively unharmed. Some of these act by interfering with the growth of the weed and are often based on plant hormones. Weed control through herbicide is made more difficult when the weeds become resistant to the herbicide. Solutions include:

- Cover crops (especially those with allelopathic properties) that out-compete weeds or inhibit their regeneration.

- Multiple herbicides, in combination or in rotation

- Strains genetically engineered for herbicide tolerance

- Locally adapted strains that tolerate or out-compete weeds

- Tilling

- Ground cover such as mulch or plastic

- Manual removal

- Mowing

- Grazing

- Burning

Terracing

Terrace rice fields in Yunnan Province, China

In agriculture, a terrace is a leveled section of a hilly cultivated area, designed as a method of soil conservation to slow or prevent the rapid surface runoff of irrigation water. Often such land is formed into multiple terraces, giving a stepped appearance. The human landscapes of rice cultivation in terraces that follow the natural contours of the escarpments like contour ploughing is a classic feature of the island of Bali and the Banaue Rice Terraces in Banaue, Ifugao, Philippines. In Peru, the Inca made use of otherwise unusable slopes by drystone walling to create terraces.

Rice Paddies

A paddy field is a flooded parcel of arable land used for growing rice and other semiaquatic crops. Paddy fields are a typical feature of rice-growing countries of east and southeast Asia including Malaysia, China, Sri Lanka, Myanmar, Thailand, Korea, Japan, Vietnam, Taiwan, Indonesia, India, and the Philippines. They are also found in other rice-growing regions such as Piedmont (Italy), the Camargue (France) and the Artibonite Valley (Haiti). They can occur naturally along rivers or marshes, or can be constructed, even on hillsides. They require large water quantities for irrigation, much of it from flooding. It gives an environment favourable to the strain of rice being grown, and is hostile to many species of weeds. As the only draft animal species which is comfortable in wetlands, the water buffalo is in widespread use in Asian rice paddies.

Paddy-based rice-farming has been practiced in Korea since ancient times. A pit-house at the Daecheon-ni archaeological site yielded carbonized rice grains and radiocarbon dates indicating that rice cultivation may have begun as early as the Middle Jeulmun Pottery Period (c. 3500-2000 BC) in the Korean Peninsula. The earliest rice cultivation there may have used dry-fields instead of paddies.

The earliest Mumun features were usually located in naturally swampy, low-lying narrow gulleys and fed by local streams. Some Mumun paddies in flat areas were made of a series of squares and rectangles separated by bunds approximately 10 cm in height, while terraced paddies consisted of long irregularly shapes that followed natural contours of the land at various levels.

Like today's, Mumun period rice farmers used terracing, bunds, canals and small reservoirs. Some paddy-farming techniques of the Middle Mumun (c. 850-550 BC) can be interpreted from the well-preserved wooden tools excavated from archaeological rice paddies at the Majeon-ni Site. However, iron tools for paddy-farming were not introduced until sometime after 200 BC. The spatial scale of individual paddies, and thus entire paddy-fields, increased with the regular use of iron tools in the Three Kingdoms of Korea Period (c. AD 300/400-668).

A recent development in the intensive production of rice is System of Rice Intensification (SRI). Developed in 1983 by the French Jesuit Father Henri de Laulanié in Madagascar, by 2013 the number of smallholder farmers using SRI had grown to between 4 and 5 million.

Aquaculture

Aquaculture is the cultivation of the natural products of water (fish, shellfish, algae, seaweed and other aquatic organisms). Intensive aquaculture takes place on land using tanks, ponds or other controlled systems or in the ocean, using cages.

Sustainable Intensive Farming

Sustainable intensive farming practises have been developed to slow the deterioration of agricultural land and even regenerate soil health and ecosystem services, while still offering high yields. Most of these developments fall in the category of organic farming, or the integration of organic and conventional agriculture.

"Organic systems and the practices that make them effective are being picked up more and more by conventional agriculture and will become the foundation for future farming systems. They won't be called organic, because they'll still use some chemicals and still use some fertilizers, but they'll function much more like today's organic systems than today's conventional systems."

Dr. Charles Benbrook Executive director US House Agriculture Subcommittee Director Agricultural Board - National Academy Sciences (FMR)

The System of Crop Intensification (SCI) was born out of research primarily at Cornell University and smallholder farms in India on SRI. It uses the SRI concepts and methods for rice and applies them to crops like wheat, sugarcane, finger millet, and others. It can be 100% organic, or integrated with reduced conventional inputs.

Holistic management is a systems thinking approach that was originally developed for reversing desertification. Holistic planned grazing is similar to rotational grazing but differs in that it more explicitly provides a framework for adapting to four basic ecosystem processes: the water cycle, the mineral cycle including the carbon cycle, energy flow and community dynamics (the relationship between organisms in an ecosystem) as equal in importance to livestock production and social welfare. By intensively managing the behavior and movement of livestock, holistic planned grazing simultaneously increases stocking rates and restores grazing land.

Pasture cropping plants grain crops directly into grassland without first applying herbicides. The perennial grasses form a living mulch understory to the grain crop, eliminating the need to plant cover crops after harvest. The pasture is intensively grazed both before and after grain production using holistic planned grazing. This intensive system yields equivalent farmer profits (partly from increased livestock forage) while building new topsoil and sequestering up to 33 tons of CO_2/ha/year.

The Twelve Aprils grazing program for dairy production, developed in partnership with USDA-SARE, is similar to pasture cropping, but the crops planted into the perennial pasture are forage crops for dairy herds. This system improves milk production and is more sustainable than confinement dairy production.

Integrated Multi-Trophic Aquaculture (IMTA) is an example of a holistic approach. IMTA is a practice in which the by-products (wastes) from one species are recycled to become inputs (fertilizers, food) for another. Fed aquaculture (e.g. fish, shrimp) is combined with inorganic extractive (e.g. seaweed) and organic extractive (e.g. shellfish) aquaculture to create balanced systems for environmental sustainability (biomitigation), economic stability (product diversification and risk reduction) and social acceptability (better management practices).

Biointensive agriculture focuses on maximizing efficiency such as per unit area, energy input and water input. Agroforestry combines agriculture and orchard/forestry technologies to create more integrated, diverse, productive, profitable, healthy and sustainable land-use systems.

Intercropping can increase yields or reduce inputs and thus represents (potentially sustainable) agricultural intensification. However, while total yield per acre is often increased dramatically, yields of any single crop often diminish. There are also challenges to farmers relying on farming equipment optimized for monoculture, often resulting in increased labor inputs.

Vertical farming is intensive crop production on a large scale in urban centers in multi-story, artificially-lit structures that uses far less inputs and produces fewer environmental impacts.

An integrated farming system is a progressive biologically integrated sustainable agriculture system such as IMTA or Zero waste agriculture whose implementation requires exacting knowledge of the interactions of multiple species and whose benefits include sustainability and increased profitability. Elements of this integration can include:

- Intentionally introducing flowering plants into agricultural ecosystems to increase pollen-and nectar-resources required by natural enemies of insect pests

- Using crop rotation and cover crops to suppress nematodes in potatoes

Challenges

The challenges and issues of industrial agriculture for society, for the industrial agriculture sector, for the individual farm, and for animal rights include the costs and benefits of both current practices and proposed changes to those practices. This is a continuation of thousands of years of invention in feeding ever growing populations.

[W]hen hunter-gatherers with growing populations depleted the stocks of game and wild foods across the Near East, they were forced to introduce agriculture. But agriculture brought much longer hours of work and a less rich diet than hunter-gatherers enjoyed. Further population growth among shifting slash-and-burn farmers led to shorter fallow periods, falling yields and soil erosion. Plowing and fertilizers were introduced to deal with these problems - but once again involved longer hours of work and degradation of soil resources(Boserup, The Conditions of Agricultural Growth, Allen and Unwin, 1965, expanded and updated in Population and Technology, Blackwell, 1980.).

While the point of industrial agriculture is to profitably supply the world at the lowest cost, industrial methods have significant side effects. Further, industrial agriculture is not an indivisible whole, but instead is composed of multiple elements, each of which can be modified in response to market conditions, government regulation and further innovation and has its own side-effects. Various interest groups reach different conclusions on the subject.

Benefits

Population Growth

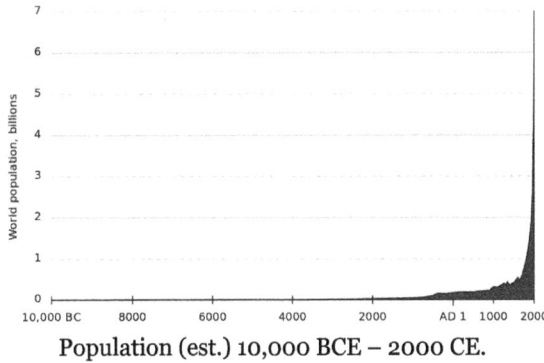

Population (est.) 10,000 BCE – 2000 CE.

Very roughly:

- 30,000 years ago hunter-gatherer behavior fed 6 million people

- 3,000 years ago primitive agriculture fed 60 million people

- 300 years ago intensive agriculture fed 600 million people

- Today **industrial agriculture** attempts to feed 6 billion people

Year	World	Africa	Asia	Europe	Central & South America	North America*	Ocea-nia	Notes
8000 BCE	8 000							
1000 BCE	50 000							
500 BCE	100 000							
1 CE	200,000 plus							
1000	310 000							
1750	791 000	106 000	502 000	163 000	16 000	2 000	2 000	
1800	978 000	107 000	635 000	203 000	24 000	7 000	2 000	
1850	1 262 000	111 000	809 000	276 000	38 000	26 000	2 000	
1900	1 650 000	133 000	947 000	408 000	74 000	82 000	6 000	

Estimated world population at various dates, in **thousands**

1950	2 518 629	221 214	1 398 488	547 403	167 097	171 616	12 812	
1955	2 755 823	246 746	1 541 947	575 184	190 797	186 884	14 265	
1960	2 981 659	277 398	1 674 336	601 401	209 303	204 152	15 888	
1965	3 334 874	313 744	1 899 424	634 026	250 452	219 570	17 657	
1970	3 692 492	357 283	2 143 118	655 855	284 856	231 937	19 443	
1975	4 068 109	408 160	2 397 512	675 542	321 906	243 425	21 564	
1980	4 434 682	469 618	2 632 335	692 431	361 401	256 068	22 828	
1985	4 830 979	541 814	2 887 552	706 009	401 469	269 456	24 678	
1990	5 263 593	622 443	3 167 807	721 582	441 525	283 549	26 687	
1995	5 674 380	707 462	3 430 052	727 405	481 099	299 438	28 924	
2000	6 070 581	795 671	3 679 737	727 986	520 229	315 915	31 043	
2005	6 453 628	887 964	3 917 508	724 722	558 281	332 156	32 998**	

An example of industrial agriculture providing cheap and plentiful food is the U.S.'s "most successful program of agricultural development of any country in the world". Between 1930 and 2000 U.S. agricultural productivity (output divided by all inputs) rose by an average of about 2 percent annually causing food prices to decrease. "The percentage of U.S. disposable income spent on food prepared at home decreased, from 22 percent as late as 1950 to 7 percent by the end of the century."

Liabilities

Environment

Industrial agriculture uses huge amounts of water, energy, and industrial chemicals; increasing pollution in the arable land, usable water and atmosphere. Herbicides, insecticides and fertilizers are accumulating in ground and surface waters. "Many of the negative effects of industrial agriculture are remote from fields and farms. Nitrogen compounds from the Midwest, for example, travel down the Mississippi to degrade coastal fisheries in the Gulf of Mexico. But other adverse effects are showing up within agricultural production systems -- for example, the rapidly developing resistance among pests is rendering our arsenal of herbicides and insecticides increasingly ineffective.". Agrochemicals and monoculture have been implicated in Colony Collapse Disorder, in which the individual members of bee colonies disappear. Agricultural production is highly dependent on bees to pollinate many varieties of fruits and vegetables.

Social

A study done for the US. Office of Technology Assessment conducted by the UC Davis Macrosocial Accounting Project concluded that industrial agriculture is associated with substantial deterioration of human living conditions in nearby rural communities.

Intensive Animal Farming

Intensive animal farming or industrial livestock production, also called factory farming by opponents of the practice, is a modern form of intensive farming that refers to the keeping of livestock, such as cattle, poultry (including in "battery cages") and fish at higher stocking densities than is usually the case with other forms of animal agriculture—a practice typical in industrial farming by agribusinesses. The main products of this industry are meat, milk and eggs for human consumption. There are issues regarding whether factory farming is sustainable and ethical.

Confinement at high stocking density is one part of a systematic effort to produce the highest output at the lowest cost by relying on economies of scale, modern machinery, biotechnology, and global trade. There are differences in the way factory farming techniques are practiced around the world. There is a continuing debate over the benefits, risks and ethical questions of factory farming. The issues include the efficiency of food production; animal welfare; whether it is essential for feeding the growing global population; and the environmental impact (e.g. pollution) and health risks.

History

The practice of industrial animal agriculture is a relatively recent development in the history of agriculture, and the result of scientific discoveries and technological advances. Innovations in agriculture beginning in the late 19th century generally parallel developments in mass production in other industries that characterized the latter part of the Industrial Revolution. The discovery of vitamins and their role in animal nutrition, in the first two decades of the 20th century, led to vitamin supplements, which allowed chickens to be raised indoors. The discovery of antibiotics and vaccines facilitated raising livestock in larger numbers by reducing disease. Chemicals developed for use in World War II gave rise to synthetic pesticides. Developments in shipping networks and technology have made long-distance distribution of agricultural produce feasible.

Agricultural production across the world doubled four times between 1820 and 1975 (1820 to 1920; 1920 to 1950; 1950 to 1965; and 1965 to 1975) to feed a global population of one billion human beings in 1800 and 6.5 billion in 2002. During the same period, the number of people involved in farming dropped as the process became more automated. In the 1930s, 24 percent of the American population worked in agriculture compared to 1.5 percent in 2002; in 1940, each farm worker supplied 11 consumers, whereas in 2002, each worker supplied 90 consumers.

According to the BBC, the era factory farming per se in Britain began in 1947 when a new Agriculture Act granted subsidies to farmers to encourage greater output by introducing new technology, in order to reduce Britain's reliance on imported meat. The United Nations writes that "intensification of animal production was seen as a way of providing food security." In 1966, the United States, United Kingdom and other industrialized nations, commenced factory farming of beef and dairy cattle and domestic pigs. From its American and West European heartland factory farming became globalised in the later years of the 20th century and is still expanding and replacing traditional practices of stock rearing in an increasing number of countries. In 1990 factory farming accounted for 30% of world meat production and by 2005 this had risen to 40%.

Contemporary Animal Production

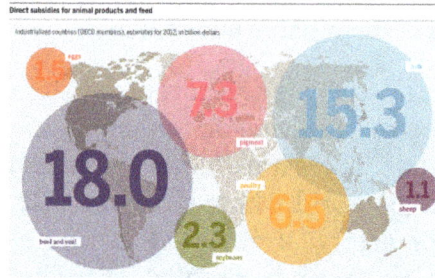

Sum of developed countries' livestock and feed subsidies

Factory farms hold large numbers of animals, typically cows, pigs, turkeys, or chickens, often indoors, typically at high densities. The aim of the operation is to produce large quantities of meat, eggs, or milk at the lowest possible cost. Food is supplied in place. Methods employed to maintain health and improve production may include some combination of disinfectants, antimicrobial agents, anthelmintics, hormones and vaccines; protein, mineral and vitamin supplements; frequent health inspections; biosecurity; climate-controlled facilities and other measures. Physical restraints, e.g. fences or creeps, are used to control movement or actions regarded as undesirable. Breeding programs are used to produce animals more suited to the confined conditions and able to provide a consistent food product.

Intensive production of livestock and poultry is widespread in developed nations. For 2002-2003, FAO estimates of industrial production as a percentage of global production were 7 percent for beef and veal, 0.8 percent for sheep and goat meat, 42 percent for pork, and 67 percent for poultry meat. Industrial production was estimated to account for 39 percent of the sum of global production of these meats and 50 percent of total egg production. In the U.S., according to its National Pork Producers Council, 80 million of its 95 million pigs slaughtered each year are reared in industrial settings.

Chickens

In the United States, chickens were raised primarily on family farms until 1965. Originally, the primary value in poultry was eggs, and meat was considered a byproduct of egg production. Its supply was less than the demand, and poultry was expensive. Except in hot weather, eggs can be shipped and stored without refrigeration for some time before going bad; this was important in the days before widespread refrigeration.

A commercial chicken house with open sides raising broiler pullets for meat

Farm flocks tended to be small because the hens largely fed themselves through foraging, with some supplementation of grain, scraps, and waste products from other farm ventures. Such feedstuffs were in limited supply, especially in the winter, and this tended to regulate the size of the farm flocks. Soon after poultry keeping gained the attention of agricultural researchers (around 1896), improvements in nutrition and management made poultry keeping more profitable and businesslike.

Prior to about 1910, chicken was served primarily on special occasions or Sunday dinner. Poultry was shipped live or killed, plucked, and packed on ice (but not eviscerated). The "whole, ready-to-cook broiler" was not popular until the 1950s, when end-to-end refrigeration and sanitary practices gave consumers more confidence. Before this, poultry were often cleaned by the neighborhood butcher, though cleaning poultry at home was a commonplace kitchen skill.

Two kinds of poultry were generally used: broilers or "spring chickens"; young male chickens, a byproduct of the egg industry, which were sold when still young and tender (generally under 3 pounds live weight), and "stewing hens", also a byproduct of the egg industry, which were old hens past their prime for laying.

Hens in Brazil

The major milestone in 20th century poultry production was the discovery of vitamin D, which made it possible to keep chickens in confinement year-round. Before this, chickens did not thrive during the winter (due to lack of sunlight), and egg production, incubation, and meat production in the off-season were all very difficult, making poultry a seasonal and expensive proposition. Year-round production lowered costs, especially for broilers.

At the same time, egg production was increased by scientific breeding. After a few false starts, (such as the Maine Experiment Station's failure at improving egg production) success was shown by Professor Dryden at the Oregon Experiment Station.

Improvements in production and quality were accompanied by lower labor require-ments. In the 1930s through the early 1950s, 1,500 hens was considered to be a full-time job for a farm family. In the late 1950s, egg prices had fallen so dramatically that farmers typically tripled the number of hens they kept, putting three hens into what had been a single-bird cage or converting their floor-confinement houses from a single deck of roosts to triple-decker roosts. Not long after this, prices fell still further and large numbers of egg farmers left the business.

Robert Plamondon reports that the last family chicken farm in his part of Oregon, Rex Farms, had 30,000 layers and survived into the 1990s. But the standard laying house of the current operators is around 125,000 hens.

This fall in profitability was accompanied by a general fall in prices to the consumer, allowing poultry and eggs to lose their status as luxury foods.

The vertical integration of the egg and poultry industries was a late development, oc-curring after all the major technological changes had been in place for years (including the development of modern broiler rearing techniques, the adoption of the Cornish Cross broiler, the use of laying cages, etc.).

By the late 1950s, poultry production had changed dramatically. Large farms and pack-ing plants could grow birds by the tens of thousands. Chickens could be sent to slaugh-terhouses for butchering and processing into prepackaged commercial products to be frozen or shipped fresh to markets or wholesalers. Meat-type chickens currently grow to market weight in six to seven weeks, whereas only fifty years ago it took three times as long. This is due to genetic selection and nutritional modifications (but not the use of growth hormones, which are illegal for use in poultry in the US and many other countries). Once a meat consumed only occasionally, the common availability and low-er cost has made chicken a common meat product within developed nations. Growing concerns over the cholesterol content of red meat in the 1980s and 1990s further re-sulted in increased consumption of chicken.

Today, eggs are produced on large egg ranches on which environmental parameters are well controlled. Chickens are exposed to artificial light cycles to stimulate egg produc-tion year-round. In addition, it is a common practice to induce molting through careful manipulation of light and the amount of food they receive in order to further increase egg size and production.

On average, a chicken lays one egg a day, but not on every day of the year. This varies with the breed and time of year. In 1900, average egg production was 83 eggs per hen per year. In 2000, it was well over 300. In the United States, laying hens are butchered after their second egg laying season. In Europe, they are generally butchered after a single season. The laying period begins when the hen is about 18–20 weeks old (depending on breed and season). Males of the egg-type breeds have little commercial value at any age, and all those not used for breeding (roughly fifty percent of all egg-type chickens) are killed soon after hatching. The old hens also have little commercial value. Thus, the main sources of poultry meat 100 years ago (spring chickens and stewing hens) have both been entirely supplanted by meat-type broiler chickens.

Some believe that the "deadly H5N1 strain of bird flu is essentially a problem of industrial poultry practices". On the other hand, according to the CDC article *H5N1 Outbreaks and Enzootic Influenza* by Robert G. Webster et al.:

Transmission of highly pathogenic H5N1 from domestic poultry back to migratory waterfowl in western China has increased the geographic spread. The spread of H5N1 and its likely reintroduction to domestic poultry increase the need for good agricultural vaccines. In fact, the root cause of the continuing H5N1 pandemic threat may be the way the pathogenicity of H5N1 viruses is masked by cocirculating influenza viruses or bad agricultural vaccines.

Webster explains:

If you use a good vaccine you can prevent the transmission within poultry and to humans. But if they have been using vaccines now [in China] for several years, why is there so much bird flu? There is bad vaccine that stops the disease in the bird but the bird goes on pooping out virus and maintaining it and changing it. And I think this is what is going on in China. It has to be. Either there is not enough vaccine being used or there is substandard vaccine being used. Probably both. It's not just China. We can't blame China for substandard vaccines. I think there are substandard vaccines for influenza in poultry all over the world.

In response to the same concerns, Reuters reports Hong Kong infectious disease expert Lo Wing-lok saying that "The issue of vaccines has to take top priority", and Julie Hall, in charge of the WHO's outbreak response in China, saying that China's vaccinations might be "masking" the virus. The BBC reported that Wendy Barclay, a virologist at the University of Reading, UK, said:

The Chinese have made a vaccine based on reverse genetics made with H5N1 antigens, and they have been using it. There has been a lot of criticism of what they have done, because they have protected their chickens against death from this virus but the chickens still get infected; and then you get drift – the virus mutates in response to the antibodies – and now we have a situation where we have five or six "flavours" of H5N1 out there.

Keeping wild birds away from domestic birds is known to be key in the fight against H5N1. Caging (no free range poultry) is one way. Providing wild birds with restored wetlands so they naturally choose nonlivestock areas is another way that helps accomplish this. Political forces are increasingly demanding the selection of one, the other, or both based on nonscientific reasons.

Pigs

Intensive piggeries (or hog lots) are a type of concentrated animal feeding operation specialized for the raising of domestic pigs up to slaughterweight. In this system of pig production grower pigs are housed indoors in group-housing or straw-lined sheds, whilst pregnant sows are confined in sow stalls (gestation crates) and give birth in farrowing crates.

The use of sow stalls (gestation crates) has resulted in lower production costs, however, this practice has led to more significant animal welfare concerns. Many of the world's largest producers of pigs (U.S. and Canada) use sow stalls, but some nations (e.g. the UK) and some US States (e.g. Florida and Arizona) have banned them.

Intensive piggeries are generally large warehouse-like buildings. Indoor pig systems allow the pig's condition to be monitored, ensuring minimum fatalities and increased productivity. Buildings are ventilated and their temperature regulated. Most domestic pig varieties are susceptible to heat stress, and all pigs lack sweat glands and cannot cool themselves. Pigs have a limited tolerance to high temperatures and heat stress can lead to death. Maintaining a more specific temperature within the pig-tolerance range also maximizes growth and growth to feed ratio. In an intensive operation pigs will lack access to a wallow (mud), which is their natural cooling mechanism. Intensive piggeries control temperature through ventilation or drip water systems (dropping water to cool the system).

Pigs are naturally omnivorous and are generally fed a combination of grains and protein sources (soybeans, or meat and bone meal). Larger intensive pig farms may be surrounded by farmland where feed-grain crops are grown. Alternatively, piggeries are reliant on the grains industry. Pig feed may be bought packaged or mixed on-site. The intensive piggery system, where pigs are confined in individual stalls, allows each pig to be allotted a portion of feed. The individual feeding system also facilitates individual medication of pigs through feed. This has more significance to intensive farming methods, as the close proximity to other animals enables diseases to spread more rapidly. To prevent disease spreading and encourage growth, drug programs such as antibiotics, vitamins, hormones and other supplements are preemptively administered.

Indoor systems, especially stalls and pens (i.e. 'dry,' not straw-lined systems) allow for the easy collection of waste. In an indoor intensive pig farm, manure can be managed through a lagoon system or other waste-management system. However, odor remains a problem which is difficult to manage.

The way animals are housed in intensive systems varies. Breeding sows will spend the bulk of their time in sow stalls (also called gestation crates) during pregnancy or farrowing crates, with litter, until market.

Piglets often receive range of treatments including castration, tail docking to reduce tail biting, teeth clipped (to reduce injuring their mother's nipples and prevent later tusk growth) and their ears notched to assist identification. Treatments are usually made without pain killers. Weak runts may be slain shortly after birth.

Piglets also may be weaned and removed from the sows at between two and five weeks old and placed in sheds. However, grower pigs - which comprise the bulk of the herd - are usually housed in alternative indoor housing, such as batch pens. During pregnancy, the use of a stall may be preferred as it facilitates feed-management and growth control. It also prevents pig aggression (e.g. tail biting, ear biting, vulva biting, food stealing). Group pens generally require higher stockmanship skills. Such pens will usually not contain straw or other material. Alternatively, a straw-lined shed may house a larger group (i.e. not batched) in age groups.

Many countries have introduced laws to regulate treatment of farmed animals. In the USA, the federal Humane Slaughter Act requires pigs to be stunned before slaughter, although compliance and enforcement is questioned..

Cattle

Cattle, are domesticated ungulates, a member of the family Bovidae, in the subfamily Bovinae, and descended from the aurochs (*Bos primigenius*). They are raised as livestock for meat (called beef and veal), dairy products (milk), leather and as draught animals (pulling carts, plows and the like). In some countries, such as India, they are honored in religious ceremonies and revered. As of 2009–2010 it is estimated that there are 1.3–1.4 billion head of cattle in the world.

Cattle are often raised by allowing herds to graze on the grasses of large tracts of rangeland called ranches. Raising cattle in this manner allows the productive use of land that might be unsuitable for growing crops. The most common interactions with cattle involve daily feeding, cleaning and milking. Many routine husbandry practices involve ear tagging, dehorning, loading, medical operations, vaccinations and hoof care, as well as training for agricultural shows and preparations. There are also some cultural differences in working with cattle - the cattle husbandry of Fulani men rests on behavioural techniques, whereas in Europe cattle are controlled primarily by physical means like fences.

Once cattle obtain an entry-level weight, about 650 pounds (290 kg), they are transferred from the range to a feedlot to be fed a specialized animal feed which consists of corn byproducts (derived from ethanol production), barley, and other grains as well as alfalfa and cottonseed meal. The feed also contains premixes composed of microingredients such as vitamins, minerals, chemical preservatives, antibiotics, fermentation products, and other essential ingredients that are purchased from premix companies, usually in sacked form, for blending into commercial rations. Because of the availability of these products, a farmer using their own grain can formulate their own rations and be assured the animals are getting the recommended levels of minerals and vitamins. Cattle in the UK are mostly grass fed with the occasional extra such as a mineral lick or feed.

Breeders can utilise cattle husbandry to reduce M. bovis infection susceptibility by selective breeding and maintaining herd health to avoid concurrent disease. Cattle are farmed for beef, veal, dairy, leather and they are sometimes used simply to maintain grassland for wildlife - for example, in Epping Forest, England. They are often used in some of the most wild places for livestock. Depending on the breed, cattle can survive on hill grazing, heaths, marshes, moors and semi desert. Modern cows are more commercial than older breeds and having become more specialised are less versatile. For this reason many smaller farmers still favour old breeds, such as the dairy breed of cattle Jersey.

There are many potential impacts on human health due to the modern cattle industrial agriculture system. There are concerns surrounding the antibiotics and growth hormones used, increased E. Coli contamination, higher saturated fat contents in the meat because of the feed, and also environmental concerns.

As of 2010, in the U.S. 766,350 producers participate in raising beef. The beef industry is segmented with the bulk of the producers participating in raising beef calves. Beef calves are generally raised in small herds, with over 90% of the herds having less than 100 head of cattle. Fewer producers participate in the finishing phase which often occurs in a feedlot, but nonetheless there are 82,170 feedlots in the United States.

Aquaculture

Aquaculture is the cultivation of the natural produce of water (fish, shellfish, algae and other aquatic organisms). The term is distinguished from fishing by the idea of active human effort in maintaining or increasing the number of organisms involved, as opposed to simply taking them from the wild. Subsets of aquaculture include Mariculture (aquaculture in the ocean); Algaculture (the production of kelp/seaweed and other algae); Fish farming (the raising of catfish, tilapia and milkfish in freshwater and brackish ponds or salmon in marine ponds); and the growing of cultured pearls. Extensive aquaculture is based on local photosynthetical production while intensive aquaculture is based on fish fed with an external food supply.

Aquaculture has been used since ancient times and can be found in many cultures. Aquaculture was used in China c. 2500 BC. When the waters lowered after river floods, some fishes, mainly carp, were held in artificial lakes. Their brood were later fed using nymphs and silkworm feces, while the fish themselves were eaten as a source of protein. The Hawaiian people practiced aquaculture by constructing fish ponds. A remarkable example from ancient Hawaii is the construction of a fish pond, dating from at least 1,000 years ago, at Alekoko. The Japanese practiced cultivation of seaweed by providing bamboo poles and, later, nets and oyster shells to serve as anchoring surfaces for spores. The Romans often bred fish in ponds.

The practice of aquaculture gained prevalence in Europe during the Middle Ages, since fish were scarce and thus expensive. However, improvements in transportation during the 19th century made fish easily available and inexpensive, even in inland areas, causing a decline in the practice. The first North American fish hatchery was constructed on Dildo Island, Newfoundland Canada in 1889, it was the largest and most advanced in the world.

Americans were rarely involved in aquaculture until the late 20th century, but California residents harvested wild kelp and made legal efforts to manage the supply starting c. 1900, later even producing it as a wartime resource. (Peter Neushul, Seaweed for War: California's World War I kelp industry, Technology and Culture 30 (July 1989), 561–583)

In contrast to agriculture, the rise of aquaculture is a contemporary phenomenon. According to professor Carlos M. Duarte About 430 (97%) of the aquatic species presently in culture have been domesticated since the start of the 20th century, and an estimated 106 aquatic species have been domesticated over the past decade. The domestication of an aquatic species typically involves about a decade of scientific research. Current success in the domestication of aquatic species results from the 20th century rise of knowledge on the basic biology of aquatic species and the lessons learned from past success and failure. The stagnation in the world's fisheries and overexploitation of 20 to 30% of marine fish species have provided additional impetus to domesticate marine species, just as overexploitation of land animals provided the impetus for the early domestication of land species.

In the 1960s, the price of fish began to climb, as wild fish capture rates peaked and the human population continued to rise. Today, commercial aquaculture exists on an unprecedented, huge scale. In the 1980s, open-netcage salmon farming also expanded; this particular type of aquaculture technology remains a minor part of the production of farmed finfish worldwide, but possible negative impacts on wild stocks, which have come into question since the late 1990s, have caused it to become a major cause of controversy.

In 2003, the total world production of fisheries product was 132.2 million tonnes of which aquaculture contributed 41.9 million tonnes or about 31% of the total world production. The growth rate of worldwide aquaculture is very rapid (greater than 10% per year for most species) while the contribution to the total from wild fisheries has been essentially flat for the last decade.

In the US, approximately 90% of all shrimp consumed are farmed and imported. In recent years salmon aquaculture has become a major export in southern Chile, especially in Puerto Montt and Quellón, Chile's fastest-growing city.

Farmed fish are kept in concentrations never seen in the wild, e.g. 50,000 fish in a 2-acre (8,100 m²) area, with each fish occupying less room than the average bathtub. This can cause several forms of pollution. Packed tightly, fish rub against each other and the sides of their cages, damaging their fins and tails and becoming sickened with various diseases and infections.

Some species of sea lice have been noted to target farmed coho and farmed Atlantic salmon specifically. Such parasites may have an effect on nearby wild fish. For these reasons, aquaculture operators frequently need to use strong drugs to keep the fish alive (but many fish still die prematurely at rates of up to 30%) and these drugs inevitably enter the environment.

The lice and pathogen problems of the 1990s facilitated the development of current treatment methods for sea lice and pathogens. These developments reduced the stress from parasite/pathogen problems. However, being in an ocean environment, the transfer of disease organisms from the wild fish to the aquaculture fish is an ever-present risk factor.

The very large number of fish kept long-term in a single location produces a significant amount of condensed feces, often contaminated with drugs, which again affect local waterways. However, these effects appear to be local to the actual fish farm site and may be minimal to non-measurable in high current sites.

Integrated Multi-trophic Aquaculture

Integrated Multi-Trophic Aquaculture (IMTA) is a practice in which the by-products (wastes) from one species are recycled to become inputs (fertilizers, food) for another. Fed aquaculture (e.g. fish, shrimp) is combined with inorganic extractive (e.g. seaweed) and organic extractive (e.g. shellfish) aquaculture to create balanced systems for environmental sustainability (biomitigation), economic stability (product diversification and risk reduction) and social acceptability (better management practices).

"Multi-Trophic" refers to the incorporation of species from different trophic or nutritional levels in the same system. This is one potential distinction from the age-old practice of aquatic polyculture, which could simply be the co-culture of different fish species from the same trophic level. In this case, these organisms may all share the same biological and chemical processes, with few synergistic benefits, which could potentially lead to significant shifts in the ecosystem. Some traditional polyculture systems may, in fact, incorporate a greater diversity of species, occupying several niches, as extensive cultures (low intensity, low management) within the same pond. The "Integrated" in IMTA refers to the more intensive cultivation of the different species in proximity of each other, connected by nutrient and energy transfer through water, but not necessarily right at the same location.

Ideally, the biological and chemical processes in an IMTA system should balance. This is achieved through the appropriate selection and proportions of different species providing different ecosystem functions. The co-cultured species should be more than just biofilters; they should also be harvestable crops of commercial value. A working IMTA system should result in greater production for the overall system, based on mutual benefits to the co-cultured species and improved ecosystem health, even if the individual production of some of the species is lower compared to what could be reached in monoculture practices over a short term period.

Sometimes the more general term "Integrated Aquaculture" is used to describe the integration of monocultures through water transfer between organisms. For all intents and purposes however, the terms "IMTA" and "integrated aquaculture" differ primarily in their degree of descriptiveness. These terms are sometimes interchanged. Aquaponics, fractionated aquaculture, IAAS (integrated agriculture-aquaculture systems), IPUAS (integrated peri-urban-aquaculture systems), and IFAS (integrated fisheries-aquaculture systems) may also be considered variations of the IMTA concept.

Shrimp

A shrimp farm is an aquaculture business for the cultivation of marine shrimp or prawns for human consumption. Commercial shrimp farming began in the 1970s, and production grew steeply, particularly to match the market demands of the USA, Japan and Western Europe. The total global production of farmed shrimp reached more than 1.6 million tonnes in 2003, representing a value of nearly 9 Billion US$. About 75% of farmed shrimp is produced in Asia, in particular in China and Thailand. The other 25% is produced mainly in Latin America, where Brazil is the largest producer. The largest exporting nation is Thailand.

Shrimp farming has moved from China to Southeast Asia into a meat packing industry. Technological advances have led to growing shrimp at ever higher densities, and broodstock is shipped worldwide. Virtually all farmed shrimp are penaeids (i.e., of the family Penaeidae), and just two species of shrimp—the *Penaeus vannamei* (Pacific white shrimp) and the *Penaeus monodon* (giant tiger prawn)—account for roughly 80% of all farmed shrimp. These industrial monocultures are very susceptible to diseases, which have caused several regional wipe-outs of farm shrimp populations. Increasing ecological problems, repeated disease outbreaks, and pressure and criticism from both NGOs and consumer countries led to changes in the industry in the late 1990s and generally stronger regulation by governments.

Regulation

In various jurisdictions, intensive animal production of some kinds is subject to regulation for environmental protection. In the United States, a CAFO (Concentrated Animal Feeding Operation) that discharges or proposes to discharge waste requires a permit and implementation of a plan for management of manure nutrients, contaminants, wastewater, etc., as applicable, to meet requirements pursuant to the federal Clean Water Act. Some data on regulatory compliance and enforcement are available. In 2000, the US Environmental Protection Agency published 5-year and 1-year data on environmental performance of 32 industries, with data for the livestock industry being derived mostly from inspections of CAFOs. The data pertain to inspections and enforcement mostly under the Clean Water Act, but also under the Clean Air Act and Resource Conservation and Recovery Act. Of the 32 industries, livestock production was among the top seven for environmental performance over the 5-year period, and was one of the top two in the final year of that period, where good environmental performance is indicated by a low ratio of enforcement orders to inspections. The five-year and final-year ratios of enforcement/inspections for the livestock industry were 0.05 and 0.01, respectively. Also in the final year, the livestock industry was one of the two leaders among the 32 industries in terms of having the lowest percentage of facilities with violations. In Canada, intensive livestock operations are subject to provincial regulation, with definitions of regulated entities varying among provinces. Examples include Intensive Livestock Operations (Saskatchewan), Confined Feeding Operations (Alberta), Feedlots (British Columbia), High-density Permanent Outdoor Confinement Areas (Ontario) and Feedlots or Parcs d'Engraissement (Manitoba). In Canada, intensive animal production, like other agricultural sectors, is also subject to various other federal and provincial requirements.

In the United States, farmed animals are excluded by half of all state animal cruelty laws including the federal Animal Welfare Act. The 28-hour law, enacted in 1873 and amended in 1994 states that when animals are being transported for slaughter, the vehicle must stop every 28 hours and the animals must be let out for exercise, food, and water. The United States Department of Agriculture claims that the law does not apply to birds. The Humane Methods of Livestock Slaughter Act is similarly limited. Originally passed in 1958, the Act requires that livestock be stunned into unconsciousness prior to slaughter. This Act also excludes birds, who make up more than 90 percent of the animals slaughtered for food, as well as rabbits and fish. Individual states all have their own animal cruelty statutes; however many states have a provision to exempt standard agricultural practices.

In the United States there is a growing movement to mitigate the worst abuses by regulating factory farming. In Ohio animal welfare organizations reached a negotiated settlement with farm organizations while in California Proposition 2, Standards for Confining Farm Animals, an initiated law was approved by voters in 2008. Regulations have been enacted in other states and plans are underway for referendum and lobbying campaigns in other states.

An action plan has been proposed by the USDA in February 2009, called the Utilization of Manure and Other Agricultural and Industrial Byproducts. This program's goal is to protect the environment and human and animal health by using manure in a safe and effective manner. In order for this to happen, several actions need to be taken and these four components include: • Improving the Usability of Manure Nutrients through More Effective Animal Nutrition and Management • Maximizing the Value of Manure through Improved Collection, Storage, and Treatment Options • Utilizing Manure in Integrated Farming Systems to Improve Profitability and Protect Soil, Water, and Air Quality • Using Manure and Other Agricultural Byproducts as a Renewable Energy Source

In 2012 Australia's largest supermarket chain, Coles, announced that as of January 1, 2013, they will stop selling company branded pork and eggs from animals kept in factory farms. The nation's other dominant supermarket chain, Woolworths, has already begun phasing out factory farmed animal products. All of Woolworth's house brand eggs are now cage-free, and by mid-2013 all of their pork will come from farmers who operate stall-free farms.

Controversies and Criticisms

Advocates of factory farming claim that factory farming has led to the betterment of housing, nutrition, and disease control over the last twenty years, while opponents claim that it harms the environment, creates health risks, and abuses animals.

Animal Welfare

Animal welfare impacts of factory farming can include:

- Close confinement systems (cages, crates) or lifetime confinement in indoor sheds
- Discomfort and injuries caused by inappropriate flooring and housing
- Restriction or prevention of normal exercise and most of natural foraging or exploratory behaviour
- Restriction or prevention of natural maternal nesting behaviour
- Lack of daylight or fresh air and poor air quality in animal sheds
- Social stress and injuries caused by overcrowding
- Health problems caused by extreme selective breeding and management for fast growth and high productivity
- Reduced lifetime (longevity) of breeding animals (dairy cows, breeding sows)
- Fast-spreading infections encouraged by crowding and stress in intensive conditions
- Debeaking (beak trimming or shortening) in the poultry and egg industry to avoid pecking in overcrowded quarters

Confinement and overcrowding of animals results in a lack of exercise and natural lo-comotory behavior, which weakens their bones and muscles. An intensive poultry farm provides the optimum conditions for viral mutation and transmission – thousands of birds crowded together in a closed, warm, and dusty environment is highly conducive to the transmission of a contagious disease. Selecting generations of birds for their fast-er growth rates and higher meat yields has left birds' immune systems less able to cope with infections and there is a high degree of genetic uniformity in the population, mak-ing the spread of disease more likely. Further intensification of the industry has been suggested by some as the solution to avian flu, on the rationale that keeping birds in-doors will prevent contamination. However, this relies on perfect, fail-safe biosecurity – and such measures are near impossible to implement. Movement between farms by people, materials, and vehicles poses a threat and breaches in biosecurity are possible. Intensive farming may be creating highly virulent avian flu strains. With the frequent flow of goods within and between countries, the potential for disease spread is high.

Confinement and overcrowding of animals' environment presents the risk of contami-nation of the meat from viruses and bacteria. Feedlot animals reside in crowded condi-tions and often spend their time standing in their own waste. A dairy farm with 2,500 cows may produce as much waste as a city of 411,000 people, and unlike a city in which human waste ends up at a sewage treatment plant, livestock waste is not treated. As a result, feedlot animals have the potential of exposure to various viruses and bacteria via the manure and urine in their environment. Furthermore, the animals often have re-sidual manure on their bodies when they go to slaughter. Sometimes, even "free-range" animals are mutilated without the use of painkillers.

Depending on the kind of system involved, prevention and control of disease in inten-sive animal farming commonly use (where appropriate) several of biosecurity, sanita-tion, surveillance, vaccinations, antibiotics, various measures for control of parasites and other pests, preconditioning, low-stress management, and removal of infected an-imals. According to a February 2011 FDA report, nearly 29 million pounds of antimi-crobials were sold in 2009 for both therapeutic and non-therapeutic use for all farm animal species. The Union of Concerned Scientists estimates that 70% of that amount is for non-therapeutic use.

The large concentration of animals, animal waste, and the potential for dead animals in a small space poses ethical issues. It is recognized that some techniques used to sustain intensive agriculture can be cruel to animals such as mutilation. As awareness of the problems of intensive techniques has grown, there have been some efforts by govern-ments and industry to remove inappropriate techniques.

On some farms, chicks may be debeaked when very young, causing pain and shock. Confining hens and pigs in crates no larger than the animal itself may lead to physical problems such as osteoporosis and joint pain, and psychological problems including boredom, depression, and frustration, as shown by repetitive or self-destructive actions. In the UK, the Farm Animal Welfare Council was set up by the government to act as an independent advisor on animal welfare in 1979 and expresses its policy as five freedoms: from hunger & thirst; from discomfort; from pain, injury or disease; to express normal behavior; from fear and distress.

Interior of a gestational sow barn

There are differences around the world as to which practices are accepted and there continue to be changes in regulations with animal welfare being a strong driver for increased regulation. For example, the EU is bringing in further regulation to set maximum stocking densities for meat chickens by 2010, where the UK Animal Welfare Minister commented, "The welfare of meat chickens is a major concern to people throughout the European Union. This agreement sends a strong message to the rest of the world that we care about animal welfare."

Factory farming is greatly debated throughout Australia, with many people disagreeing with the methods and ways in which the animals in factory farms are treated. Animals are often under stress from being kept in confined spaces and will attack each other. In an effort to prevent injury leading to infection, their beaks, tails and teeth are removed. Many piglets will die of shock after having their teeth and tails removed, because painkilling medicines are not used in these operations. Others say that factory farms are a great way to gain space, with animals such as chickens being kept in spaces smaller than an A4 page.

Less cruel methods of factory farming are still preferable. For example, in the UK, debeaking of chickens is deprecated, but it is recognized that it is a method of last resort, seen as better than allowing vicious fighting and ultimately cannibalism. Between 60 and 70 percent of six million breeding sows in the U.S. are confined during pregnancy, and for most of their adult lives, in 2 by 7 ft (0.61 by 2.13 m) gestation crates. According to pork producers and many veterinarians, sows will fight if housed in pens. The largest pork producer in the U.S. said in January 2007 that it will phase out gestation crates by 2017. They are being phased out in the European Union, with a ban effective in 2013 after the fourth week of pregnancy. With the evolution of factory farming, there has been a growing awareness of the issues amongst the wider public, not least due to the efforts of animal rights and welfare campaigners. As a result, gestation crates, one of the more contentious practices, are the subject of laws in the U.S., Europe and around the world to phase out their use as a result of pressure to adopt less confined practices.

Human Health Impact

According to the U.S. Centers for Disease Control and Prevention (CDC), farms on which animals are intensively reared can cause adverse health reactions in farm workers. Workers may develop acute and chronic lung disease, musculoskeletal injuries, and may catch infections that transmit from animals to human beings (such as tuberculosis).

Pesticides are used to control organisms which are considered harmful and they save farmers money by preventing product losses to pests. In the US, about a quarter of pesticides used are used in houses, yards, parks, golf courses, and swimming pools and about 70% are used in agriculture. However, pesticides can make their way into consumers' bodies which can cause health problems. One source of this is bioaccumulation in animals raised on factory farms.

"Studies have discovered an increase in respiratory, neurobehavioral, and mental illnesses among the residents of communities next to factory farms."

The CDC writes that chemical, bacterial, and viral compounds from animal waste may travel in the soil and water. Residents near such farms report problems such as unpleasant smell, flies and adverse health effects.

The CDC has identified a number of pollutants associated with the discharge of animal waste into rivers and lakes, and into the air. The use of antibiotics may create antibiotic-resistant pathogens; parasites, bacteria, and viruses may be spread; ammonia, nitrogen, and phosphorus can reduce oxygen in surface waters and contaminate drinking water; pesticides and hormones may cause hormone-related changes in fish; animal feed and feathers may stunt the growth of desirable plants in surface waters and provide nutrients to disease-causing micro-organisms; trace elements such as arsenic and copper, which are harmful to human health, may contaminate surface waters.

Intensive farming may make the evolution and spread of harmful diseases easier. Many communicable animal diseases spread rapidly through densely spaced populations of animals and crowding makes genetic reassortment more likely. However, small family farms are more likely to introduce bird diseases and more frequent association with people into the mix, as happened in the 2009 flu pandemic

In the European Union, growth hormones are banned on the basis that there is no way of determining a safe level. The UK has stated that in the event of the EU raising the ban at some future date, to comply with a precautionary approach, it would only consider the introduction of specific hormones, proven on a case by case basis. In 1998, the European Union banned feeding animals antibiotics that were found to be valuable for human health. Furthermore, in 2006 the European Union banned all drugs for livestock that were used for growth promotion purposes. As a result of these bans, the levels of antibiotic resistance in animal products and within the human population showed a decrease.

The various techniques of factory farming have been associated with a number of European incidents where public health has been threatened or large numbers of animals have had to be slaughtered to deal with disease. Where disease breaks out, it may spread more quickly, not only due to the concentrations of animals, but because modern approaches tend to distribute animals more widely. The international trade in animal products increases the risk of global transmission of virulent diseases such as swine fever, BSE, foot and mouth and bird flu.

In the United States, the use of antibiotics in livestock is still prevalent. The FDA reports that 80 percent of all antibiotics sold in 2009 were administered to livestock animals, and that many of these antibiotics are identical or closely related to drugs used for treating illnesses in humans. Consequently, many of these drugs are losing their effectiveness on humans, and the total healthcare costs associated with drug-resistant bacterial infections in the United States are between $16.6 billion and $26 billion annually.

Methicillin-resistant Staphylococcus aureus (MRSA) has been identified in pigs and humans raising concerns about the role of pigs as reservoirs of MRSA for human infection. One study found that 20% of pig farmers in the United States and Canada in 2007 harbored MRSA. A second study revealed that 81% of Dutch pig farms had pigs with MRSA and 39% of animals at slaughter carried the bug were all of the infections were resistant to tetracycline and many were resistant to other antimicrobials. A more recent study found that MRSA ST398 isolates were less susceptible to tiamulin, an antimicrobial used in agriculture, than other MRSA or methicillin susceptible *S. aureus*. Cases of MRSA have increased in livestock animals. CC398 is a new clone of MRSA that has emerged in animals and is found in intensively reared production animals (primarily pigs, but also cattle and poultry), where it can be transmitted to humans. Although dangerous to humans, CC398 is often asymptomatic in food-producing animals.

A 2011 nationwide study reported nearly half of the meat and poultry sold in U.S. grocery stores — 47 percent — was contaminated with S. aureus, and more than half of those bacteria — 52 percent — were resistant to at least three classes of antibiotics. Although Staph should be killed with proper cooking, it may still pose a risk to consumers through improper food handling and cross-contamination in the kitchen. The senior author of the study said, "The fact that drug-resistant S. aureus was so prevalent, and likely came from the food animals themselves, is troubling, and demands attention to how antibiotics are used in food-animal production today."

In April 2009, lawmakers in the Mexican state of Veracruz accused large-scale hog and poultry operations of being breeding grounds of a pandemic swine flu, although they did not present scientific evidence to support their claim. A swine flu which quickly killed more than 100 infected persons in that area, appears to have begun in the vicinity of a Smithfield subsidiary pig CAFO (concentrated animal feeding operation).

Environmental Impact

Intensive factory farming has grown to become the biggest threat to the global environment through the loss of ecosystem services and global warming. It is a major driver to global environmental degradation. The process in which feed needs to be grown for animal use only is often grown using intensive methods which involve a significant amount of fertiliser and pesticides. This sometimes results in the pollution of water, soil and air by agrochemicals and manure waste, and use of limited resources such as water and energy at unsustainable rates.

Industrial production of pigs and poultry is an important source of GHG emissions and is predicted to become more so. On intensive pig farms, the animals are generally kept on concrete with slats or grates for the manure to drain through. The manure is usually stored in slurry form (slurry is a liquid mixture of urine and feces). During storage on farm, slurry emits methane and when manure is spread on fields it emits nitrous oxide and causes nitrogen pollution of land and water. Poultry manure from factory farms emits high levels of nitrous oxide and ammonia.

Large quantities and concentrations of waste are produced. Air quality and groundwater are at risk when animal waste is improperly recycled.

Environmental impacts of factory farming include:

- Deforestation for animal feed production
- Unsustainable pressure on land for production of high-protein/high-energy animal feed
- Pesticide, herbicide and fertilizer manufacture and use for feed production
- Unsustainable use of water for feed-crops, including groundwater extraction
- Pollution of soil, water and air by nitrogen and phosphorus from fertiliser used for feed-crops and from manure
- Land degradation (reduced fertility, soil compaction, increased salinity, desertification)
- Loss of biodiversity due to eutrophication, acidification, pesticides and herbicides
- Worldwide reduction of genetic diversity of livestock and loss of traditional breeds
- Species extinctions due to livestock-related habitat destruction (especially feed-cropping)

Labor

Small farmers are often absorbed into factory farm operations, acting as contract growers for the industrial facilities. In the case of poultry contract growers, farmers are required to make costly investments in construction of sheds to house the birds, buy required feed and drugs - often settling for slim profit margins, or even losses.

Market Concentration

The major concentration of the industry occurs at the slaughter and meat processing phase, with only four companies slaughtering and processing 81 percent of cows, 73 percent of sheep, 57 percent of pigs and 50 percent of chickens. This concentration at the slaughter phase may be in large part due to regulatory barriers that may make it financially difficult for small slaughter plants to be built, maintained or remain in business. Factory farming may be no more beneficial to livestock producers than traditional farming because it appears to contribute to overproduction that drives down prices. Through "forward contracts" and "marketing agreements", meatpackers are able to set the price of livestock long before they are ready for production. These strategies often cause farmers to lose money, as half of all U.S. family farming operations did in 2007.

In 1967, there were one million pig farms in America; as of 2002, there were 114,000.

Many of the nation's livestock producers would like to market livestock directly to consumers but with limited USDA inspected slaughter facilities, livestock grown locally can not typically be slaughtered and processed locally.

Demonstrations

From 2011 to 2014 each year between 15,000 and 30,000 people gathered under the theme *We are fed up!* in Berlin to protest against industrial livestock production.

Managed Intensive Rotational Grazing

In agriculture, Managed intensive rotational grazing (MIRG), also known as cell grazing, mob grazing and holistic managed planned grazing, describes a variety of closely related systems of forage use in which ruminant and non-ruminant herds and/or flocks are regularly and systematically moved to fresh rested areas with the intent to maximize the quality and quantity of forage growth.

One primary goal of MIRG is to have a vegetative cover over all grazed areas at all times, and to prevent the complete removal of all vegetation from the grazed areas ("bare dirt")

MIRG can be used with cattle, sheep, goats, pigs, chickens, turkeys, ducks and other animals. The herds graze one portion of pasture, or a paddock, while allowing the others to recover. The length of time a paddock is grazed will depend on the size of the herd and the size of the paddock. Resting grazed lands allows the vegetation to renew energy reserves, rebuild shoot systems, and deepen root systems, with the result being long-term maximum biomass production. MIRG is especially effective because grazers do better on the more tender younger plant stems. MIRG also leave parasites behind to die off minimizing or eliminating the need for de-wormers. Pasture systems alone can allow grazers to meet their energy requirements, and with the increased productivity of MIRG systems, the grazers obtain the majority of their nutritional needs without the supplemental feed sources that are required in continuous grazing systems.

One key element of this style of animal husbandry is that either each grazed area must contain all elements needed for the animals (water source, for instance) or the feed or water source must be moved each time the animals are moved. Having fixed feeding or watering stations defeats the rotational aspect, leading to degradation of the ground around the water supply or feed supply if additional feed is provided to the animals. Special care must be taken to ensure that high use areas do not become areas where mud, parasites or diseases are spread or communicated

Herd Health and Welfare

Animal Health Risks

Bloat is a common problem in grazing systems for ruminants, although not for pigs or poultry, that if left untreated can lead to animal death. This problem occurs when foam producing compounds in plants are digested by cows, causing foam to form in the rumen of the animal and ultimately prohibiting animals from expelling gas. The risk of bloat can be mitigated by seeding non-bloating legumes with the grasses. Animals are especially susceptible to bloat if they are moved to new pasture sources when they are particularly hungry. It is therefore important to ensure that the herd is eating enough at the end of a rotation when forage will be more scarce, limiting the potential for animals to gorge themselves when turned onto new paddocks.

Several problems are related to shade in pasture areas. Although shade provides relief from heat and reduces the risk of heat stress, animals tend to congregate in these areas which leads to nutrient loading, uneven grazing, and potential soil erosion. Taller shade trees move the shade area around as the day progresses minimizing this problem.

Animal Health Benefits and Animal Welfare

Herd health benefits arise from animals having access to both space and fresh air. Freedom of movement within a paddock results in increased physical fitness, which limits the potential for injuries and abrasion, and reduces the potential of exposure to high levels of harmful disease-causing microorganisms and insects.

In a concentrated animal feeding operation (CAFO), it is considered normal for a large number of animals to continuously occupy a very small area. The aisles that animals use to move around are consequently constantly coated with a moist layer of manure and urine from the many animals, leading to ailments such as foot rot due to the constant wet exposure. The manure and urine is usually just scraped off into gutters below slatted surfaces, and the surfaces and gutters are rarely washed completely clean, so molds, bacteria, and insects can grow and thrive in the potentially infectious waste. Feeding areas in a CAFO are also rarely stripped bare and washed with a disinfectant, as this would reduce the time available for animals to eat continuously, so molds and bacteria are also able to become established where the animals eat. The close confinement and lack of general environmental cleanliness leads to easier spread of infection and increased sickness, requiring the regular feeding of antibiotics to keep the confined animals healthy but which also leads to antibiotic resistance for bacteria constantly present in the CAFO.

By comparison, with managed grazing, the animals are able to spread out and exist in a natural environment more suited for their natural growth and development. As the animals move to a new paddock, wastes are left behind and allowed to decay without the animals nearby. The animals experience less disease without the need for regular antibiotic dosing, and fewer foot ailments.

Weed Control

In general, a well managed rotational grazing system has rather low pasture weed establishment because the majority of niches are already filled with established forage species, making it hard for competing weeds to emerge and become established. The use of multiple species in the co-grazing helps to minimize weeds. Established forage plants in rotational grazing pasture systems are healthy and unstressed due to the "rest" period, enhancing the competitive advantage of the forage. Additionally, in comparison to cash grain crop production, many plants which would be considered weeds are not problematic in perennial pasture. Many of these plants are actually nutritious to grazers and control of these plants is therefore not necessary in management intensive rotational systems. However, certain species such as thistles and various other weeds, are indigestible and possibly poisonous to grazers. These plant species will not be grazed by the herd and can be recognized for their prevalence in pasture systems.

A key step in managing weeds in any pasture system is identification. Once the undesired species in a pasture system are identified, an integrated approach of management can be implemented to control weed populations. It is important to recognize that no single approach to weed management will result in weed free pastures; therefore, various cultural, mechanical, and chemical control methods can be combined in an integrated weed management plan. Cultural controls include: avoiding spreading manure contaminated with weed seeds, cleaning equipment after working in weed infested areas, and managing weed problems in fencerows and other areas near pastures.

Mechanical controls such as repeated mowing, clipping, and hand weeding can also be used to effectively manage weed infestations by weakening the plant. These methods should be implemented when weed flower buds are closed or just starting to open to prevent seed production. Although these methods for managing weeds greatly reduces reliance on herbicides, weed problems may still persist in managed grazing systems and the use of herbicides may become necessary. Use of herbicides may restrict the use of a pasture for some length of time, depending on the type and amount of the chemical used. Frequently, weeds in pasture systems are patchy and therefore spot treatment of herbicides may be used as a least cost method of chemical control.

Nutrient Availability and Soil Fertility

If pasture systems are seeded with more than 40% legumes, commercial nitrogen fertilization is unnecessary for adequate plant growth. Legumes are able to fix atmospheric nitrogen, thus providing nitrogen for themselves and surrounding plants. Although grazers remove nutrient sources from the pasture system when they feed on forage sources, the majority of the nutrients consumed by the herd are returned to the pasture system through manure. At a relatively high stocking rate, or high ratio of animals per hectare, manure will be evenly distributed across the pasture system. The nutrient content in these manure sources should be adequate to meet plant requirements, making commercial fertilization unnecessary. Management intensive rotational grazing systems are often associated with increased soil fertility which arises because manure is a rich source of organic matter that increases the health of soil. In addition, these pasture system are less susceptible to erosion because the land base has continuous ground cover throughout the year.

High levels of fertilizers entering waterways are a pertinent environmental concern associated with agricultural systems. However, management intensive rotational grazing systems effectively reduce the amount of nutrients that move off-farm which have the potential to cause environmental degradation. These systems are fertilized with on-farm sources, and are less prone to leaching as compared to commercial fertilizers. Additionally, the system is less prone to excess nutrient fertilization, so the majority of nutrients put into the system by manure sources are utilized for plant growth. Permanent pasture systems also have deeper, well established forage root systems which are more efficient at taking up nutrients from within the soil profile.

Socio-cultural-economic Considerations

Although milk yields are often lower in MIRG systems, net farm income per cow is often greater as compared to confinement operations. This is due to the additional costs associated with herd health and purchased feeds are greatly reduced in management intensive rotational grazing systems. Additionally, a transition to management intensive rotational grazing is associated with low start-up and maintenance costs. Another consideration is that while production per cow is less, the amount of cows per acre on the pasture can increase. The net effect is more productivity per acre at less cost.

The main costs associated with transitioning to management intensive rotational grazing are purchasing fencing, fencers, and water supply materials. If a pasture was continuously grazed in the past, likely capital has already been invested in fencing and a fencer system. Cost savings to graziers can also be recognized when one considers that many of the costs associated with livestock operations are transmitted to the grazers. For example, the grazers actively harvest their own sources of food for the portion of the year where grazing is possible. This translates into lower costs for feed production and harvesting, which are fuel intensive endeavors. MIRG systems rely on the grazers to produce fertilizer sources via their excretion. There is also no need for collection, storage, transportation, and application of manure, which are also all fuel intensive. Additionally, external fertilizer use contributes to other costs such as labor, purchasing costs.

It can also be demonstrated that management intensive rotational grazing system also result in time savings because the majority of work which might otherwise require human labor is transmitted to the herd.

Environmental Considerations

Many pastures undergoing MIRG are less susceptible to soil erosion and are associated with higher soil fertility than continuously grazed pastures, depending on the skill of the manager and the management system he is using. As a result, the paddocks require fewer commercial inputs, which have been associated with negative environmental impacts. In addition, because these systems tend to be more resilient and stable they are more capable of responding to changing environmental conditions and perturbations while not compromising productivity.

Human Nutrition

Animals raised on pasture have shown major differences in the nutritional quality of the products they produce for human consumption.

Criticism

Managed intensive rotational grazing paints a wide brush over many different managed grazing systems. Managers have found that rotational grazing systems can work for diverse management purposes, but scientific experiments have demonstrated that rotational grazing systems do not always necessarily work for specific ecological purposes. This controversy stems from two main categorical differences in rotational grazing, prescribed management and adaptive management. The performance of rangeland grazing strategies are similarly constrained by several ecological variables establishing that differences among them are dependent on the effectiveness of those management models. Depending on the management model, plant production has been shown to be equal or greater in continuous compared to rotational grazing in 87% of the experiments.

Intensive Crop Farming

Intensive crop farming is a modern form of intensive farming that refers to the industrialized production of crops. Intensive crop farming's methods include innovation in agricultural machinery, farming methods, genetic engineering technology, techniques for achieving economies of scale in production, the creation of new markets for consumption, patent protection of genetic information, and global trade. These methods are widespread in developed nations.

The practice of industrial agriculture is a relatively recent development in the history of agriculture, and the result of scientific discoveries and technological advances. Innovations in agriculture beginning in the late 19th century generally parallel developments in mass production in other industries that characterized the latter part of the Industrial Revolution. The identification of nitrogen and phosphorus as critical factors in plant growth led to the manufacture of synthetic fertilizers, making more intensive uses of farmland for crop production possible.

Similarly, the discovery of vitamins and their role in animal nutrition, in the first two decades of the 20th century, led to vitamin supplements, which in the 1920s allowed certain livestock to be raised indoors, reducing their exposure to adverse natural elements. The discovery of antibiotics and vaccines facilitated raising livestock in larger numbers by reducing disease. Chemicals developed for use in World War II gave rise to synthetic pesticides. Developments in shipping networks and technology have made long-distance distribution of produce feasible.

Crops

Features

- large scale – hundreds or thousands of acres of a single crop (much more than can be absorbed into the local or regional market);

- monoculture – large areas of a single crop, often raised from year to year on the same land, or with little crop rotation;

- agrichemicals – reliance on imported, synthetic fertilizers and pesticides to provide nutrients and to mitigate pests and diseases, these applied on a regular schedule

- hybrid seed – use of specialized hybrids designed to favor large scale distribution (e.g. ability to ripen off the vine, to withstand shipping and handling);

- genetically engineered crops – use of genetically modified varieties designed for large scale production (e.g. ability to withstand selected herbicides);

- large scale irrigation – heavy water use, and in some cases, growing of crops in otherwise unsuitable regions by extreme use of water (e.g. rice paddies on arid land).

- high mechanization – automated machinery sustain and harvest crops.

Criticism

Critics of intensively farmed crops cite a wide range of concerns. On the food quality front, it is held by critics that quality is reduced when crops are bred and grown primarily for cosmetic and shipping characteristics. Environmentally, industrial farming of crops is claimed to be responsible for loss of biodiversity, degradation of soil quality, soil erosion, food toxicity (pesticide residues) and pollution (through agrichemical build-ups and runoff, and use of fossil fuels for agrichemical manufacture and for farm machinery and long-distance distribution).

History

The projects within the Green Revolution spread technologies that had already existed, but had not been widely used outside of industrialized nations. These technologies included pesticides, irrigation projects, and synthetic nitrogen fertilizer.

The novel technological development of the Green Revolution was the production of what some referred to as "miracle seeds." Scientists created strains of maize, wheat, and rice that are generally referred to as HYVs or "high-yielding varieties." HYVs have an increased nitrogen-absorbing potential compared to other varieties. Since cereals that absorbed extra nitrogen would typically lodge, or fall over before harvest, semi-dwarfing genes were bred into their genomes. Norin 10 wheat, a variety developed by Orville Vogel from Japanese dwarf wheat varieties, was instrumental in developing Green Revolution wheat cultivars. IR8, the first widely implemented HYV rice to be developed by IRRI, was created through a cross between an Indonesian variety named "Peta" and a Chinese variety named "Dee Geo Woo Gen."

With the availability of molecular genetics in Arabidopsis and rice the mutant genes responsible (*reduced height(rht), gibberellin insensitive (gai1)* and *slender rice (slr1)*) have been cloned and identified as cellular signalling components of gibberellic acid, a phytohormone involved in regulating stem growth via its effect on cell division. Stem growth in the mutant background is significantly reduced leading to the dwarf phenotype. Photosynthetic investment in the stem is reduced dramatically as the shorter plants are inherently more stable mechanically. Assimilates become redirected to grain production, amplifying in particular the effect of chemical fertilisers on commercial yield.

HYVs significantly outperform traditional varieties in the presence of adequate irrigation, pesticides, and fertilizers. In the absence of these inputs, traditional varieties may outperform HYVs. One criticism of HYVs is that they were developed as F1 hybrids, meaning they need to be purchased by a farmer every season rather than saved from previous seasons, thus increasing a farmer's cost of production.

Examples

Wheat (Modern Management Techniques)

Wheat is a grass that is cultivated worldwide. Globally, it is the most important human food grain and ranks second in total production as a cereal crop behind maize; the third being rice. Wheat and barley were the first cereals known to have been domesticated. Cultivation and repeated harvesting and sowing of the grains of wild grasses led to the domestication of wheat through selection of mutant forms with tough ears which remained intact during harvesting, and larger grains. Because of the loss of seed dispersal mechanisms, domesticated wheats have limited capacity to propagate in the wild.

Agricultural cultivation using horse collar leveraged plows (3000 years ago) increased cereal grain productivity yields, as did the use of seed drills which replaced broadcasting sowing of seed in the 18th century. Yields of wheat continued to increase, as new land came under cultivation and with improved agricultural husbandry involving the use of fertilizers, threshing machines and reaping machines (the 'combine harvester'), tractor-draw cultivators and planters, and better varieties. With population growth rates falling, while yields continue to rise, the area devoted to wheat may now begin to decline for the first time in modern human history.

Organic wheat typically halves yield attainable but costs less as there are no fertiliser and pesticide costs. Seed costs are typically higher, however, and arguably labour and machinery costs are higher as the organic crop, and more importantly the whole rotation and cropping on such a farm, is more difficult to manage correctly.

While winter wheat lies dormant during a winter freeze, wheat normally requires between 110 and 130 days between planting and harvest, depending upon climate, seed type, and soil conditions. Crop management decisions require the knowledge of stage of development of the crop. In particular, spring fertilizers applications, herbicides, fungicides, growth regulators are typically applied at specific stages of plant development. For example, current recommendations often indicate the second application of nitrogen be done when the ear (not visible at this stage) is about 1 cm in size (Z31 on Zadoks scale).

Maize (Mechanical Harvesting)

Maize was planted by the Native Americans in hills, in a complex system known to some as the Three Sisters: beans used the corn plant for support, and squashes provided ground cover to stop weeds. This method was replaced by single species hill planting where each hill 60–120 cm (2–4 ft) apart was planted with 3 or 4 seeds, a method still used by home gardeners. A later technique was *checked corn* where hills were placed 40 inches (1,000 mm) apart in each direction, allowing cultivators to run through the field in two directions. In more arid lands this was altered and seeds were planted in the bottom of 10–12 cm (4–5 in) deep furrows to collect water. Modern technique plants maize in rows which allows for cultivation while the plant is young, although the hill technique is still used in the cornfields of some Native American reservations. Haudenosaunee Confederacy is what a group of Native Americans who are preparing for climate change through seed banking. Now this group is known as the Iroquois.

With a climate changing more crops are able to grow in different areas that they previously weren't able to grow in. This will open growing areas for maize.

A corn heap at the harvest site, India

In North America, fields are often planted in a two-crop rotation with a nitrogen-fixing crop, often alfalfa in cooler climates and soybeans in regions with longer summers. Sometimes a third crop, winter wheat, is added to the rotation. Fields are usually plowed each year, although no-till farming is increasing in use. Many of the maize varieties grown in the United States and Canada are hybrids. Over half of the corn area planted in the United States has been genetically modified using biotechnology to express agronomic traits such as pest resistance or herbicide resistance.

Before about World War II, most maize in North America was harvested by hand (as it still is in most of the other countries where it is grown). This often involved large numbers of workers and associated social events. Some one- and two-row mechanical pickers were in use but the corn combine was not adopted until after the War. By hand or mechanical picker, the entire ear is harvested which then requires a separate operation of a corn sheller to remove the kernels from the ear. Whole ears of corn were often stored in *corn cribs* and these whole ears are a sufficient form for some livestock feeding use. Few modern farms store maize in this manner. Most harvest the grain from the field and store it in bins. The combine with a corn head (with points and snap rolls instead of a reel) does not cut the stalk; it simply pulls the stalk down. The stalk continues downward and is crumpled into a mangled pile on the ground. The ear of corn is too large to pass through a slit in a plate and the snap rolls pull the ear of corn from the stalk so that only the ear and husk enter the machinery. The combine separates the husk and the cob, keeping only the kernels.

Soybean (Genetic Modification)

Soybeans are one of the "biotech food" crops that are being genetically modified, and GMO soybeans are being used in an increasing number of products. Monsanto Company is the world's leader in genetically modified soy for the commercial market. In 1995, Monsanto introduced "Roundup Ready" (RR) soybeans that have had a copy of a gene from the bacterium, *Agrobacterium* sp. strain CP4, inserted, by means of a gene gun, into its genome that allows the transgenic plant to survive being sprayed by this non-selective herbicide, glyphosate. Glyphosate, the active ingredient in Roundup, kills conventional soybeans. The bacterial gene is EPSP (= 5-enolpyruvyl shikimic acid-3-phosphate) synthase. Soybean also has a version of this gene, but the soybean version is sensitive to glyphosate, while the CP4 version is not.

RR soybeans allow a farmer to reduce tillage or even to sow the seed directly into an unplowed field, known as 'no-till' or conservation tillage. No-till agriculture has many advantages, greatly reducing soil erosion and creating better wildlife habitat; it also saves fossil fuels, and sequesters CO_2, a greenhouse effect gas.

In *1997*, about 8% of all soybeans cultivated for the commercial market in the United States were genetically modified. In 2006, the figure was 89%. As with other "Roundup Ready" crops, concern is expressed over damage to biodiversity. However, the RR gene has been bred into so many different soybean cultivars that the genetic modification itself has not resulted in any decline of genetic diversity.

Tomato (Hydroponics)

The largest commercial hydroponics facility in the world is Eurofresh Farms in Willcox, Arizona, which sold more than 200 million pounds of tomatoes in 2007. Eurofresh has 318 acres (1.3 km²) under glass and represents about a third of the commercial hydroponic greenhouse area in the U.S. Eurofresh does not consider their tomatoes organic, but they are pesticide-free. They are grown in rockwool with top irrigation.

Some commercial installations use no pesticides or herbicides, preferring integrated pest management techniques. There is often a price premium willingly paid by consumers for produce which is labeled "organic". Some states in the USA require soil as an essential to obtain organic certification. There are also overlapping and somewhat contradictory rules established by the US Federal Government. So some food grown with hydroponics can be certified organic. In fact, they are the cleanest plants possible because there is no environment variable and the dirt in the food supply is extremely limited. Hydroponics also saves an incredible amount of water; It uses as little as 1/20 the amount as a regular farm to produce the same amount of food. The water table can be impacted by the water use and run-off of chemicals from farms, but hydroponics may minimize impact as well as having the advantage that water use and water returns are easier to measure. This can save the farmer money by allowing reduced water use and the ability to measure consequences to the land around a farm.

The environment in a hydroponics greenhouse is tightly controlled for maximum efficiency and this new mindset is called soil-less/controlled-environment agriculture (S/CEA). With this growers can make ultra-premium foods anywhere in the world, regardless of temperature and growing seasons. Growers monitor the temperature, humidity, and pH level constantly.

Crop Rotation

Satellite image of circular crop fields in Kansas in late June 2001. Healthy, growing crops are green. Corn would be growing into leafy stalks by then. Sorghum, which resembles corn, grows more slowly and would be much smaller and therefore, (possibly) paler. Wheat is a brilliant yellow as harvest occurs in June. Fields of brown have been recently harvested and plowed under or lie fallow for the year.

Effects of crop rotation and monoculture at the Swojec Experimental Farm, Wroclaw University of Environmental and Life Sciences. In the front field, the "Norfolk" crop rotation sequence (potatoes, oats, peas, rye) is being applied; in the back field, rye has been grown for 45 years in a row.

Crop rotation is the practice of growing a series of dissimilar or different types of crops in the same area in sequenced seasons. It is done so that the soil of farms is not used to only one type of nutrient. It helps in reducing soil erosion and increases soil fertility and crop yield.

Growing the same crop in the same place for many years in a row disproportionately depletes the soil of certain nutrients. With rotation, a crop that leaches the soil of one kind of nutrient is followed during the next growing season by a dissimilar crop that returns that nutrient to the soil or draws a different ratio of nutrients. In addition, crop rotation mitigates the buildup of pathogens and pests that often occurs when one species is continuously cropped, and can also improve soil structure and fertility by increasing biomass from varied root structures.

Crop rotation is used in both conventional and organic farming systems.

History

Agriculturalists have long recognized that suitable rotations – such as planting spring crops for livestock in place of grains for human consumption – make it possible to restore or to maintain a productive soil. Middle Eastern farmers practised crop rotation in 6000 BC without understanding the chemistry, alternately planting legumes and cereals. In the Bible, chapter 25 of the Book of Leviticus instructs the Israelites to observe a "Sabbath of the Land". Every seventh year they would not till, prune or even control insects. The Roman writer, Cato the Elder (234 – 149 BC), recommended that farmers "save carefully goat, sheep, cattle, and all other dung". From the times of Charlemagne (died 814), farmers in Europe transitioned from a two-field crop rotation to a three-field crop rotation. Under a two-field rotation, half the land was planted in a year, while the other half lay fallow. Then, in the next year, the two fields were reversed.

From the end of the Middle Ages until the 20th century, Europe's farmers practised three-field rotation, dividing available lands into three parts. One section was planted in the autumn with rye or winter wheat, followed by spring oats or barley; the second section grew crops such as peas, lentils, or beans; and the third field was left fallow. The three fields were rotated in this manner so that every three years, a field would rest and be fallow. Under the two-field system, if one has a total of 600 acres (2.4 km²) of fertile land, one would only plant 300 acres. Under the new three-field rotation system, one would plant (and therefore harvest) 400 acres. But the additional crops had a more significant effect than mere quantitative productivity. Since the spring crops were mostly legumes, they increased the overall nutrition of the people of Northern Europe.

Farmers in the region of Waasland (in present-day northern Belgium) pioneered a four-field rotation in the early 16th century, and the British agriculturist Charles Townshend (1674-1738) popularised this system in the 18th century. The sequence of four crops (wheat, turnips, barley and clover), included a fodder crop and a grazing crop, allowing livestock to be bred year-round. The four-field crop rotation became a key development in the British Agricultural Revolution.

George Washington Carver (1860s - 1943) studied crop-rotation methods in the United States, teaching southern farmers to rotate soil-depleting crops like cotton with soil-enriching crops like peanuts and peas.

In the Green Revolution of the mid-20th century the traditional practice of crop rotation gave way in some parts of the world to the practice of supplementing the chemical inputs to the soil through topdressing with fertilizers, adding (for example) ammonium nitrate or urea and restoring soil pH with lime. Such practices aimed to increase yields, to prepare soil for specialist crops, and to reduce waste and inefficiency by simplifying planting and harvesting.

Crop Choice

A preliminary assessment of crop interrelationships can be found in how each crop: (1) contributes to soil organic matter (SOM) content, (2) provides for pest management, (3) manages deficient or excess nutrients, and (4) how it contributes to or controls for soil erosion.

Crop choice is often related to the goal the farmer is looking to achieve with the rotation, which could be weed management, increasing available nitrogen in the soil, controlling for erosion, or increasing soil structure and biomass, to name a few. When discussing crop rotations, crops are classified in different ways depending on what quality is being assessed: by family, by nutrient needs/benefits, and/or by profitability (i.e. cash crop versus cover crop). For example, giving adequate attention to plant family is essential to mitigating pests and pathogens. However, many farmers have success managing rotations by planning sequencing and cover crops around desirable cash crops. The following is a simplified classification based on crop quality and purpose.

Row Crops

Many crops which are critical for the market, like vegetables, are row crops (that is, grown in tight rows). While often the most profitable for farmers, these crops are more taxing on the soil Row crops typically have low biomass and shallow roots: this means the plant contributes low residue to the surrounding soil and has limited effects on structure. With much of the soil around the plant is exposed to disruption by rainfall and traffic, fields with row crops experience faster break down of organic matter by microbes, leaving fewer nutrients for future plants.

In short, while these crops may be profitable for the farm, they are nutrient depleting. Crop rotation practices exist to strike a balance between short-term profitability and long-term productivity.

Legumes

A great advantage of crop rotation comes from the interrelationship of nitrogen fixing-crops with nitrogen demanding crops. Legumes, like alfalfa and clover, collect available nitrogen from the soil in nodules on their root structure. When the plant is harvested, the biomass of uncollected roots breaks down, making the stored nitrogen available to future crops. Legumes are also a valued green manure: a crop that collects nutrients and fixes them at soil depths accessible to future crops.

In addition, legumes have heavy tap roots that burrow deep into the ground, lifting soil for better tilth and absorption of water.

Grasses and Cereals

Cereal and grasses are frequent cover crops because of the many advantages they supply to soil quality and structure. The dense and far-reaching root systems give ample structure to surrounding soil and provide significant biomass for soil organic matter.

Grasses and cereals are key in weed management as they compete with undesired plants for soil space and nutrients.

Green Manure

Green manure is a crop that is mixed into the soil. Both nitrogen-fixing legumes and nutrient scavengers, like grasses, can be used as green manure. Green manure of legumes is an excellent source of nitrogen, especially for organic systems, however, legume biomass doesn't contribute to lasting soil organic matter like grasses do.

Planning a Rotation

There are numerous factors that must be taken into consideration when planning a crop rotation. Planning an effective rotation requires weighing fixed and fluctuating production circumstances, including, but not limited to: market, farm size, labor supply, climate, soil type, growing practices, etc. Moreover, a crop rotation must consider in what condition one crop will leave the soil for the succeeding crop and how one crop can be seeded with another crop. For example, a nitrogen-fixing crop, like a legume, should always proceed a nitrogen depleting one; similarly, a low residue crop (i.e. a crop with low biomass) should be offset with a high biomass cover crop, like a mixture of grasses and legumes.

There is no limit to the number of crops that can be used in a rotation, or the amount of time a rotation takes to complete. Decisions about rotations are made years prior, seasons prior, or even at the very last minute when an opportunity to increase profits or soil quality presents itself. In short, there is no singular formula for rotation, but many considerations to take into account.

Implementation

Crop rotation systems may be enriched by the influences of other practices such as the addition of livestock and manure, intercropping or multiple cropping, and organic management low in pesticides and synthetic fertilizers.

Incorporation of Livestock

Introducing livestock makes the most efficient use of critical sod and cover crops; livestock (through manure) are able to distribute the nutrients in these crops throughout the soil rather than removing nutrients from the farm through the sale of hay. In systems where use of farm livestock would violate reservations growers or consumers may have about animal exploitation, efforts are made to surrogate this input through livestock in the soil, namely worms and microorganisms.

In Sub-Saharan Africa, as animal husbandry becomes less of a nomadic practice many herders have begun integrating crop production into their practice. This is known as mixed farming, or the practice of crop cultivation with the incorporation of raising cattle, sheep and/or goats by the same economic entity, is increasingly common. This interaction between the animal, the land and the crops are being done on a small scale all across this region. Crop residues provide animal feed, while the animals provide manure for replenishing crop nutrients and draft power. Both processes are extremely important in this region of the world as it is expensive and logistically unfeasible to transport in synthetic fertilizers and large-scale machinery. As an additional benefit, the cattle, sheep and/or goat provide milk and can act as a cash crop in the times of economic hardship.

Organic Farming

Crop rotation is a required practice in order for a farm to receive organic certification in the United States. The "Crop Rotation Practice Standard" for the National Organic Program under the U.S. Code of Federal Regulations, section §205.205, states that:

Farmers are required to implement a crop rotation that maintains or builds soil organic matter, works to control pests, manages and conserves nutrients, and protects against erosion. Producers of perennial crops that aren't rotated may utilize other practices, such as cover crops, to maintain soil health.

In addition to lowering the need for inputs by controlling for pests and weeds and increasing available nutrients, crop rotation helps organic growers increase the amount of biodiversity on their farms. Biodiversity is also a requirement of organic certification, however, there are no rules in place to regulate or reinforce this standard. Increasing the biodiversity of crops has beneficial effects on the surrounding ecosystem and can host a greater diversity of fauna, insects, and beneficial microorganism in the soil.< Some studies point to increased nutrient availability from crop rotation under organic systems compared to conventional practices as organic practices are less likely to inhibit of beneficial microbes in soil organic matter.

While multiple cropping and intercropping benefit from many of the same principals as crop rotation, they do not satisfy the requirement under the NOP.

Intercropping

Multiple cropping systems, such as intercropping or companion planting, offer more diversity and complexity within the same season or rotation, for example the three sisters. An example of companion planting is the inter-planting of corn with pole beans and vining squash or pumpkins. In this system, the beans provide nitrogen; the corn provides support for the beans and a "screen" against squash vine borer; the vining squash provides a weed suppressive canopy and discourages corn-hungry raccoons.

Double-cropping is common where two crops, typically of different species, are grown sequentially in the same growing season, or where one crop (e.g. vegetable) is grown continuously with a cover crop (e.g. wheat). This is advantageous for small farms, who often cannot afford to leave cover crops to replenish the soil for extended periods of time, as larger farms can. When multiple cropping is implemented on small farms, these systems can maximize benefits of crop rotation on available land resources.

Benefits

Agronomists describe the benefits to yield in rotated crops as "The Rotation Effect". There are many found benefits of rotation systems; however, there is no specific scientific basis for the sometimes 10-25% yield increase in a crop grown in rotation versus monoculture. The factors related to the increase are simply described as alleviation of the negative factors of monoculture cropping systems. Explanations due to improved nutrition; pest, pathogen, and weed stress reduction; and improved soil structure have been found in some cases to be correlated, but causation has not been determined for the majority of cropping systems.

Other benefits of rotation cropping systems include production cost advantages. Overall financial risks are more widely distributed over more diverse production of crops and/or livestock. Less reliance is placed on purchased inputs and over time crops can maintain production goals with fewer inputs. This in tandem with greater short and long term yields makes rotation a powerful tool for improving agricultural systems.

Soil Organic Matter

The use of different species in rotation allows for increased soil organic matter (SOM), greater soil structure, and improvement of the chemical and biological soil environment for crops. With more SOM, water infiltration and retention improves, providing increased drought tolerance and decreased erosion.

Soil organic matter is a mix of decaying material from biomass with active microorganisms. Crop rotation, by nature, increases exposure to biomass from sod, green manure, and a various other plant debris. The reduced need for intensive tillage under crop rotation allows biomass aggregation to lead to greater nutrient retention and utilization, decreasing the need for added nutrients. With tillage, disruption and oxidation of soil creates a less conducive environment for diversity and proliferation of microorganisms in the soil. These microorganisms are what make nutrients available to plants. So, where "active" soil organic matter is a key to productive soil, soil with low microbial activity provides significantly fewer nutrients to plants; this is true even though the quantity of biomass left in the soil may be the same.

Soil microorganisms also decrease pathogen and pest activity through competition. In addition, plants produce root exudates and other chemicals which manipulate their soil environment as well as their weed environment. Thus rotation allows increased yields from nutrient availability but also alleviation of allelopathy and competitive weed environments.

Carbon Sequestration

Studies have shown that crop rotations greatly increase soil organic carbon (SOC) content, the main constituent of soil organic matter. Carbon, along with hydrogen and oxygen, is a macronutrient for plants. Highly diverse rotations spanning long periods of time have shown to be even more effective in increasing SOC, while soil disturbances (e.g. from tillage) are responsible for exponential decline in SOC levels. In Brazil, conservation to no-till methods combined with intensive crop rotations has been shown an SOC sequestration rate of 0.41 tonnes per hectare per year.

In addition to enhancing crop productivity, sequestration of atmospheric carbon has great implications in reducing rates of climate change by removing carbon dioxide from the air.

Nitrogen Fixing

Rotating crops adds nutrients to the soil. Legumes, plants of the family Fabaceae, for instance, have nodules on their roots which contain nitrogen-fixing bacteria called rhizobia. It therefore makes good sense agriculturally to alternate them with cereals (family Poaceae) and other plants that require nitrates.

Pathogen and Pest Control

Crop rotation is also used to control pests and diseases that can become established in the soil over time. The changing of crops in a sequence decreases the population level of pests by (1) interrupting pest life cycles and (2) interrupting pest habitat. Plants within the same taxonomic family tend to have similar pests and pathogens. By regularly changing crops and keeping the soil occupied by cover crops instead of lying fallow, pest cycles can be broken or limited, especially cycles that benefit from overwintering in residue. For example, root-knot nematode is a serious problem for some plants in warm climates and sandy soils, where it slowly builds up to high levels in the soil, and can severely damage plant productivity by cutting off circulation from the plant roots.

Growing a crop that is not a host for root-knot nematode for one season greatly reduces the level of the nematode in the soil, thus making it possible to grow a susceptible crop the following season without needing soil fumigation.

This principle is of particular use in organic farming, where pest control must be achieved without synthetic pesticides.

Weed Management

Integrating certain crops, especially cover crops, into crop rotations is of particular value to weed management. These crops crowd out weed through competition. In addition, the sod and compost from cover crops and green manure slows the growth of what weeds are still able to make it through the soil, giving the crops further competitive advantage. By removing slowing the growth and proliferation of weeds while cover crops are cultivated, farmers greatly reduce the presence of weeds for future crops, including shallow rooted and row crops, which are less resistant to weeds. Cover crops are, therefore, considered conservation crops because they protect otherwise fallow land from becoming overrun with weeds.

This system has advantages over other common practices for weeds management, such as tillage. Tillage is meant to inhibit growth of weeds by overturning the soil; however, this has a countering effect of exposing weed seeds that may have gotten buried and burying valuable crop seeds. Under crop rotation, the number of viable seeds in the soil is reduced through the reduction of the weed population.

Preventing Soil Erosion

Crop rotation can significantly reduce the amount of soil lost from erosion by water. In areas that are highly susceptible to erosion, farm management practices such as zero and reduced tillage can be supplemented with specific crop rotation methods to reduce raindrop impact, sediment detachment, sediment transport, surface runoff, and soil loss.

Protection against soil loss is maximized with rotation methods that leave the greatest mass of crop stubble (plant residue left after harvest) on top of the soil. Stubble cover in contact with the soil minimizes erosion from water by reducing overland flow velocity, stream power, and thus the ability of the water to detach and transport sediment. Soil Erosion and Cill prevent the disruption and detachment of soil aggregates that cause macropores to block, infiltration to decline, and runoff to increase. This significantly improves the resilience of soils when subjected to periods of erosion and stress.

The effect of crop rotation on erosion control varies by climate. In regions under relatively consistent climate conditions, where annual rainfall and temperature levels are assumed, rigid crop rotations can produce sufficient plant growth and soil cover. In regions where climate conditions are less predictable, and unexpected periods of rain and drought may occur, a more flexible approach for soil cover by crop rotation is necessary. An opportunity cropping system promotes adequate soil cover under these erratic climate conditions. In an opportunity cropping system, crops are grown when soil water is adequate and there is a reliable sowing window. This form of cropping system is likely to produce better soil cover than a rigid crop rotation because crops are only sown under optimal conditions, whereas rigid systems are not necessarily sown in the best conditions available.

Crop rotations also affect the timing and length of when a field is subject to fallow. This is very important because depending on a particular region's climate, a field could be the most vulnerable to erosion when it is under fallow. Efficient fallow management is an essential part of reducing erosion in a crop rotation system. Zero tillage is a fundamental management practice that promotes crop stubble retention under longer unplanned fallows when crops cannot be planted. Such management practices that succeed in retaining suitable soil cover in areas under fallow will ultimately reduce soil loss.

Biodiversity

Increasing the biodiversity of crops has beneficial effects on the surrounding ecosystem and can host a greater diversity of fauna, insects, and beneficial microorganisms in the soil. Some studies point to increased nutrient availability from crop rotation under organic systems compared to conventional practices as organic practices are less likely to inhibit of beneficial microbes in soil organic matter, such as arbuscular mycorrhizae, which increase nutrient uptake in plants. Increasing biodiversity also increases the resilience of agro-ecological systems.

Farm Productivity

Crop rotation contributes to increased yields through improved soil nutrition. By requiring planting and harvesting of different crops at different times, more land can be farmed with the same amount of machinery and labour.

Risk Management

Different crops in the rotation can reduce the risks of adverse weather for the individual farmer.

Challenges

While crop rotation requires a great deal of planning, crop choice must respond to a number of fixed conditions (soil type, topography, climate, and irrigation) in addition to conditions that may change dramatically from year to the next (weather, market, labor supply). In this way, it is unwise to plan crops years in advance. Improper implementation of a crop rotation plan may lead to imbalances in the soil nutrient composition or a buildup of pathogens affecting a critical crop. The consequences of faulty rotation may take years to become apparent even to experienced soil scientists and can take just as long to correct.

Many challenges exist within the practices associated with crop rotation. For example, green manure from legumes can lead to an invasion of snails or slugs and the decay from green manure can occasionally suppress the growth of other crops.

Irrigation

An irrigation sprinkler watering a lawn

Irrigation canal in Osmaniye, Turkey

Irrigation is the method in which a controlled amount of water is supplied to plants at regular intervals for agriculture. It is used to assist in the growing of agricultural crops, maintenance of landscapes, and revegetation of disturbed soils in dry areas and during periods of inadequate rainfall. Additionally, irrigation also has a few other uses in crop production, which include protecting plants against frost, suppressing weed growth in grain fields and preventing soil consolidation. In contrast, agriculture that relies only on direct rainfall is referred to as rain-fed or dry land farming.

Irrigation systems are also used for dust suppression, disposal of sewage, and in mining. Irrigation is often studied together with drainage, which is the natural or artificial removal of surface and sub-surface water from a given area.

Irrigation has been a central feature of agriculture for over 5,000 years and is the product of many cultures. Historically, it was the basis for economies and societies across the globe, from Asia to the Southwestern United States.

History

Animal-powered irrigation, Upper Egypt, ca. 1846

Inside a karez tunnel at Turpan, Sinkiang

ArchaMIhceal ified as evidence of irrigation where the natural rainfall was insufficient to support crops for rainfed agriculture.

Perennial irrigation was practiced in the Mesopotamian plain whereby crops were regularly watered throughout the growing season by coaxing water through a matrix of small channels formed in the field.

irrigation in Tamil Nadu (India)

Ancient Egyptians practiced *Basin irrigation* using the flooding of the Nile to inundate land plots which had been surrounded by dykes. The flood water was held until the fertile sediment had settled before the surplus was returned to the watercourse. There is evidence of the ancient Egyptian pharaoh Amenemhet III in the twelfth dynasty (about 1800 BCE) using the natural lake of the Faiyum Oasis as a reservoir to store surpluses of water for use during the dry seasons, the lake swelled annually from flooding of the Nile.

The Ancient Nubians developed a form of irrigation by using a waterwheel-like device called a *sakia*. Irrigation began in Nubia some time between the third and second millennium BCE. It largely depended upon the flood waters that would flow through the Nile River and other rivers in what is now the Sudan.

In sub-Saharan Africa irrigation reached the Niger River region cultures and civilizations by the first or second millennium BCE and was based on wet season flooding and water harvesting.

Terrace irrigation is evidenced in pre-Columbian America, early Syria, India, and China. In the Zana Valley of the Andes Mountains in Peru, archaeologists found remains of three irrigation canals radiocarbon dated from the 4th millennium BCE, the 3rd millennium BCE and the 9th century CE. These canals are the earliest record of irrigation in the New World. Traces of a canal possibly dating from the 5th millennium BCE were found under the 4th millennium canal. Sophisticated irrigation and storage systems were developed by the Indus Valley Civilization in present-day Pakistan and North India, including the reservoirs at Girnar in 3000 BCE and an early canal irrigation system from circa 2600 BCE. Large scale agriculture was practiced and an extensive network of canals was used for the purpose of irrigation.

Ancient Persia (modern day Iran) as far back as the 6th millennium BCE, where barley was grown in areas where the natural rainfall was insufficient to support such a crop. The Qanats, developed in ancient Persia in about 800 BCE, are among the oldest known irrigation methods still in use today. They are now found in Asia, the Middle East and North Africa. The system comprises a network of vertical wells and gently sloping tunnels driven into the sides of cliffs and steep hills to tap groundwater. The noria, a water wheel with clay pots around the rim powered by the flow of the stream (or by animals where the water source was still), was first brought into use at about this time, by Roman settlers in North Africa. By 150 BCE the pots were fitted with valves to allow smoother filling as they were forced into the water.

The irrigation works of ancient Sri Lanka, the earliest dating from about 300 BCE, in the reign of King Pandukabhaya and under continuous development for the next thousand years, were one of the most complex irrigation systems of the ancient world. In addition to underground canals, the Sinhalese were the first to build completely artificial reservoirs to store water. Due to their engineering superiority in this sector, they were often called 'masters of irrigation'.Most of these irrigation systems still exist undamaged up to now, in Anuradhapura and Polonnaruwa, because of the advanced and precise engineering. The system was extensively restored and further extended during the reign of King Parakrama Bahu (1153–1186 CE).

China

The oldest known hydraulic engineers of China were Sunshu Ao (6th century BCE) of the Spring and Autumn Period and Ximen Bao (5th century BCE) of the Warring States period, both of whom worked on large irrigation projects. In the Sichuan region belonging to the State of Qin of ancient China, the Dujiangyan Irrigation System was built in 256 BCE to irrigate an enormous area of farmland that today still supplies water. By the 2nd century AD, during the Han Dynasty, the Chinese also used chain pumps that lifted water from lower elevation to higher elevation. These were powered by manual foot pedal, hydraulic waterwheels, or rotating mechanical wheels pulled by oxen. The water was used for public works of providing water for urban residential quarters and palace gardens, but mostly for irrigation of farmland canals and channels in the fields.

Korea

In 15th century Korea, the world's first rain gauge, *uryanggye* (Korean:우량계), was invented in 1441. The inventor was Jang Yeong-sil, a Korean engineer of the Joseon Dynasty, under the active direction of the king, Sejong the Great. It was installed in irrigation tanks as part of a nationwide system to measure and collect rainfall for agricultural applications. With this instrument, planners and farmers could make better use of the information gathered in the survey.

North America

In North America, the Hohokam were the only culture known to rely on irrigation canals to water their crops, and their irrigation systems supported the largest population in the Southwest by AD 1300. The Hohokam constructed an assortment of simple canals combined with weirs in their various agricultural pursuits. Between the 7th and 14th centuries, they also built and maintained extensive irrigation networks along the lower Salt and middle Gila rivers that rivaled the complexity of those used in the ancient Near East, Egypt, and China. These were constructed using relatively simple excavation tools, without the benefit of advanced engineering technologies, and achieved drops of a few feet per mile, balancing erosion and siltation. The Hohokam cultivated varieties of cotton, tobacco, maize, beans and squash, as well as harvested an assortment of wild plants. Late in the Hohokam Chronological Sequence, they also used extensive dry-farming systems, primarily to grow agave for food and fiber. Their reliance on agricultural strategies based on canal irrigation, vital in their less than hospitable desert environment and arid climate, provided the basis for the aggregation of rural populations into stable urban centers.

Present Extent

Irrigation ditch in Montour County, Pennsylvania, off Strawberry Ridge Road

In the mid-20th century, the advent of diesel and electric motors led to systems that could pump groundwater out of major aquifers faster than drainage basins could refill them. This can lead to permanent loss of aquifer capacity, decreased water quality, ground subsidence, and other problems. The future of food production in such areas as the North China Plain, the Punjab, and the Great Plains of the US is threatened by this phenomenon.

At the global scale, 2,788,000 km² (689 million acres) of fertile land was equipped with irrigation infrastructure around the year 2000. About 68% of the area equipped for irrigation is located in Asia, 17% in the America, 9% in Europe, 5% in Africa and 1% in Oceania. The largest contiguous areas of high irrigation density are found:

- In Northern India and Pakistan along the Ganges and Indus rivers

- In the Hai He, Huang He and Yangtze basins in China

- Along the Nile river in Egypt and Sudan

- In the Mississippi-Missouri river basin and in parts of California

Smaller irrigation areas are spread across almost all populated parts of the world.

Only eight years later in 2008, the scale of irrigated land increased to an estimated total of 3,245,566 km² (802 million acres), which is nearly the size of India.

Types of Irrigation

Various types of irrigation techniques differ in how the water obtained from the source is distributed within the field. In general, the goal is to supply the entire field uniformly with water, so that each plant has the amount of water it needs, neither too much nor too little.

Basin flood irrigation of wheat

Irrigation of land in Punjab, Pakistan

Surface Irrigation

In *surface* (*furrow*, *flood*, or *level basin*) irrigation systems, water moves across the surface of agricultural lands, in order to wet it and infiltrate into the soil. Surface irrigation can be subdivided into furrow, *borderstrip or basin irrigation*. It is often called *flood irrigation* when the irrigation results in flooding or near flooding of the cultivated land. Historically, this has been the most common method of irrigating agricultural land and still used in most parts of the world.

Where water levels from the irrigation source permit, the levels are controlled by dikes, usually plugged by soil. This is often seen in terraced rice fields (rice paddies), where the method is used to flood or control the level of water in each distinct field. In some cases, the water is pumped, or lifted by human or animal power to the level of the land. The field water efficiency of surface irrigation is typically lower than other forms of irrigation but has the potential for efficiencies in the range of 70% - 90% under appropriate management.

Localized Irrigation

Localized irrigation is a system where water is distributed under low pressure through a piped network, in a pre-determined pattern, and applied as a small discharge to each plant or adjacent to it. Drip irrigation, spray or micro-sprinkler irrigation and bubbler irrigation belong to this category of irrigation methods.

Impact sprinkler head

Subsurface Textile Irrigation

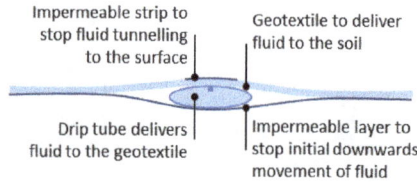

Impermeable strip to
stop fluid tunnelling
to the surface

Geotextile to deliver
fluid to the soil

Drip tube delivers
fluid to the geotextile

Impermeable layer to
stop initial downwards
movement of fluid

Diagram showing the structure of an example SSTI installation

Subsurface Textile Irrigation (SSTI) is a technology designed specifically for subsurface irrigation in all soil textures from desert sands to heavy clays. A typical subsurface textile irrigation system has an impermeable base layer (usually polyethylene or polypropylene), a drip line running along that base, a layer of geotextile on top of the drip line and, finally, a narrow impermeable layer on top of the geotextile. Unlike standard drip irrigation, the spacing of emitters in the drip pipe is not critical as the geotextile moves the water along the fabric up to 2 m from the dripper.

Drip Irrigation

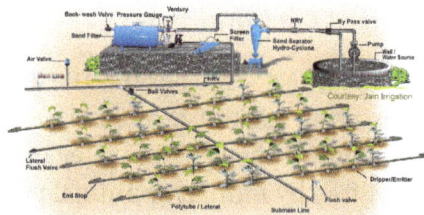

Drip irrigation layout and its parts

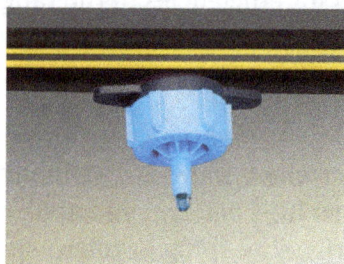

Drip irrigation – a dripper in action

Grapes in Petrolina, only made possible in this semi arid area by drip irrigation

Drip (or micro) irrigation, also known as trickle irrigation, functions as its name suggests. In this system water falls drop by drop just at the position of roots. Water is delivered at or near the root zone of plants, drop by drop. This method can be the most water-efficient method of irrigation, if managed properly, since evaporation and runoff are minimized. The field water efficiency of drip irrigation is typically in the range of 80 to 90 percent when managed correctly.

In modern agriculture, drip irrigation is often combined with plastic mulch, further reducing evaporation, and is also the means of delivery of fertilizer. The process is known as *fertigation*.

Deep percolation, where water moves below the root zone, can occur if a drip system is operated for too long or if the delivery rate is too high. Drip irrigation methods range from very high-tech and computerized to low-tech and labor-intensive. Lower water pressures are usually needed than for most other types of systems, with the exception of low energy center pivot systems and surface irrigation systems, and the system can be designed for uniformity throughout a field or for precise water delivery to individual plants in a landscape containing a mix of plant species. Although it is difficult to regulate pressure on steep slopes, pressure compensating emitters are available, so the field does not have to be level. High-tech solutions involve precisely calibrated emitters located along lines of tubing that extend from a computerized set of valves.

Irrigation Using Sprinkler Systems

Sprinkler irrigation of blueberries in Plainville, New York, United States

A traveling sprinkler at Millets Farm Centre, Oxfordshire, United Kingdom

In *sprinkler* or overhead irrigation, water is piped to one or more central locations within the field and distributed by overhead high-pressure sprinklers or guns. A system utilizing sprinklers, sprays, or guns mounted overhead on permanently installed risers is often referred to as a *solid-set* irrigation system. Higher pressure sprinklers that rotate are called *rotors* an are driven by a ball drive, gear drive, or impact mechanism. Rotors can be designed to rotate in a full or partial circle. Guns are similar to rotors, except that they generally operate at very high pressures of 40 to 130 lbf/in^2 (275 to 900 kPa) and flows of 50 to 1200 US gal/min (3 to 76 L/s), usually with nozzle diameters in the range of 0.5 to 1.9 inches (10 to 50 mm). Guns are used not only for irrigation, but also for industrial applications such as dust suppression and logging.

Sprinklers can also be mounted on moving platforms connected to the water source by a hose. Automatically moving wheeled systems known as *traveling sprinklers* may irrigate areas such as small farms, sports fields, parks, pastures, and cemeteries unattended. Most of these utilize a length of polyethylene tubing wound on a steel drum. As the tubing is wound on the drum powered by the irrigation water or a small gas engine, the sprinkler is pulled across the field. When the sprinkler arrives back at the reel the system shuts off. This type of system is known to most people as a "waterreel" traveling irrigation sprinkler and they are used extensively for dust suppression, irrigation, and land application of waste water.

Other travelers use a flat rubber hose that is dragged along behind while the sprinkler platform is pulled by a cable. These cable-type travelers are definitely old technology and their use is limited in today's modern irrigation projects.

Irrigation Using Center Pivot

Center pivot irrigation is a form of sprinkler irrigation consisting of several segments of pipe (usually galvanized steel or aluminium) joined together and supported by trusses, mounted on wheeled towers with sprinklers positioned along its length. The system moves in a circular pattern and is fed with water from the pivot point at the center of the arc. These systems are found and used in all parts of the world and allow irrigation of all types of terrain. Newer systems have drop sprinkler heads as shown in the image that follows.

Center pivot with drop sprinklers

Most center pivot systems now have drops hanging from a u-shaped pipe attached at the top of the pipe with sprinkler head that are positioned a few feet (at most) above the crop, thus limiting evaporative losses. Drops can also be used with drag hoses or bubblers that deposit the water directly on the ground between crops. Crops are often planted in a circle to conform to the center pivot. This type of system is known as LEPA (Low Energy Precision Application). Originally, most center pivots were water powered. These were replaced by hydraulic systems (*T-L Irrigation*) and electric motor driven systems (Reinke, Valley, Zimmatic). Many modern pivots feature GPS devices.

Wheel line irrigation system in Idaho, 2001

Irrigation by Lateral Move (Side Roll, Wheel Line, Wheelmove)

A *series of pipes, each with a wheel* of about 1.5 m diameter permanently affixed to its midpoint, and sprinklers along its length, are coupled together. Water is supplied at one end using a large hose. After sufficient irrigation has been applied to one strip of the field, the hose is removed, the water drained from the system, and the assembly rolled either by hand or with a purpose-built mechanism, so that the sprinklers are moved to a different position across the field. The hose is reconnected. The process is repeated in a pattern until the whole field has been irrigated.

This system is less expensive to install than a center pivot, but much more labor-intensive to operate - it does not travel automatically across the field: it applies water in a stationary strip, must be drained, and then rolled to a new strip. Most systems use 4 or 5-inch (130 mm) diameter aluminum pipe. The pipe doubles both as water transport and as an axle for rotating all the wheels. A drive system (often found near the centre of the wheel line) rotates the clamped-together pipe sections as a single axle, rolling the whole wheel line. Manual adjustment of individual wheel positions may be necessary if the system becomes misaligned.

Wheel line systems are limited in the amount of water they can carry, and limited in the height of crops that can be irrigated. One useful feature of a lateral move system is that it consists of sections that can be easily disconnected, adapting to field shape as the line is moved. They are most often used for small, rectilinear, or oddly-shaped fields, hilly or mountainous regions, or in regions where labor is inexpensive.

Sub-irrigation

Subirrigation has been used for many years in field crops in areas with high water tables. It is a method of artificially raising the water table to allow the soil to be moistened from below the plants' root zone. Often those systems are located on permanent grasslands in lowlands or river valleys and combined with drainage infrastructure. A system of pumping stations, canals, weirs and gates allows it to increase or decrease the water level in a network of ditches and thereby control the water table.

Sub-irrigation is also used in commercial greenhouse production, usually for potted plants. Water is delivered from below, absorbed upwards, and the excess collected for recycling. Typically, a solution of water and nutrients floods a container or flows through a trough for a short period of time, 10–20 minutes, and is then pumped back into a holding tank for reuse. Sub-irrigation in greenhouses requires fairly sophisticated, expensive equipment and management. Advantages are water and nutrient conservation, and labor-saving through lowered system maintenance and automation. It is similar in principle and action to subsurface basin irrigation.

Irrigation Automatically, Non-electric Using Buckets and Ropes

Besides the common manual watering by bucket, an automated, natural version of this also exists. Using plain polyester ropes combined with a prepared ground mixture can be used to water plants from a vessel filled with water.

The ground mixture would need to be made depending on the plant itself, yet would mostly consist of black potting soil, vermiculite and perlite. This system would (with certain crops) allow to save expenses as it does not consume any electricity and only little water (unlike sprinklers, water timers, etc.). However, it may only be used with certain crops (probably mostly larger crops that do not need a humid environment; perhaps e.g. paprikas).

Irrigation Using Water Condensed from Humid Air

In countries where at night, humid air sweeps the countryside.Water can be obtained from the humid air by condensation onto cold surfaces. This is for example practiced in the vineyards at Lanzarote using stones to condense water or with various fog collectors based on canvas or foil sheets.

In-ground Irrigation

Most commercial and residential irrigation systems are "in ground" systems, which means that everything is buried in the ground. With the pipes, sprinklers, emitters (drippers), and irrigation valves being hidden, it makes for a cleaner, more presentable landscape without garden hoses or other items having to be moved around manually. This does, however, create some drawbacks in the maintenance of a completely buried system.

Most irrigation systems are divided into zones. A zone is a single irrigation valve and one or a group of drippers or sprinklers that are connected by pipes or tubes. Irrigation systems are divided into zones because there is usually not enough pressure and available flow to run sprinklers for an entire yard or sports field at once. Each zone has a solenoid valve on it that is controlled via wire by an irrigation controller. The irrigation controller is either a mechanical (now the "dinosaur" type) or electrical device that signals a zone to turn on at a specific time and keeps it on for a specified amount of time. "Smart Controller" is a recent term for a controller that is capable of adjusting the watering time by itself in response to current environmental conditions. The smart controller determines current conditions by means of historic weather data for the local area, a soil moisture sensor (water potential or water content), rain sensor, or in more sophisticated systems satellite feed weather station, or a combination of these.

When a zone comes on, the water flows through the lateral lines and ultimately ends up at the irrigation emitter (drip) or sprinkler heads. Many sprinklers have pipe thread inlets on the bottom of them which allows a fitting and the pipe to be attached to them. The sprinklers are usually installed with the top of the head flush with the ground surface. When the water is pressurized, the head will pop up out of the ground and water the desired area until the valve closes and shuts off that zone. Once there is no more water pressure in the lateral line, the sprinkler head will retract back into the ground. Emitters are generally laid on the soil surface or buried a few inches to reduce evaporation losses.

Water Sources

Irrigation water can come from groundwater (extracted from springs or by using wells), from surface water (withdrawn from rivers, lakes or reservoirs) or from non-conventional sources like treated wastewater, desalinated water or drainage water. A special form of irrigation using surface water is spate irrigation, also called floodwater harvesting. In case of a flood (spate), water is diverted to normally dry river beds (wadis) using a network of dams, gates and channels and spread over large areas. The moisture stored in the soil will be used thereafter to grow crops. Spate irrigation areas are in particular located in semi-arid or arid, mountainous regions. While floodwater harvesting belongs to the accepted irrigation methods, rainwater harvesting is usually not considered as a form of irrigation. Rainwater harvesting is the collection of runoff water from roofs or unused land and the concentration of this.

Irrigation is underway by pump-enabled extraction directly from the Gumti, seen in the background, in Comilla, Bangladesh.

Around 90% of wastewater produced globally remains untreated, causing widespread water pollution, especially in low-income countries. Increasingly, agriculture uses untreated wastewater as a source of irrigation water. Cities provide lucrative markets for fresh produce, so are attractive to farmers. However, because agriculture has to compete for increasingly scarce water resources with industry and municipal users, there is often no alternative for farmers but to use water polluted with urban waste, including sewage, directly to water their crops. Significant health hazards can result from using water loaded with pathogens in this way, especially if people eat raw vegetables that have been irrigated with the polluted water. The International Water Management Institute has worked in India, Pakistan, Vietnam, Ghana, Ethiopia, Mexico and other countries on various projects aimed at assessing and reducing risks of wastewater irrigation. They advocate a 'multiple-barrier' approach to wastewater use, where farmers are encouraged to adopt various risk-reducing behaviours. These include ceasing irrigation a few days before harvesting to allow pathogens to die off in the sunlight, applying water carefully so it does not contaminate leaves likely to be eaten raw, cleaning vegetables with disinfectant or allowing fecal sludge used in farming to dry before being used as a human manure. The World Health Organization has developed guidelines for safe water use.

There are numerous benefits of using recycled water for irrigation, including the low cost (when compared to other sources, particularly in an urban area), consistency of supply (regardless of season, climatic conditions and associated water restrictions), and general consistency of quality. Irrigation of recycled wastewater is also considered as a means for plant fertilization and particularly nutrient supplementation. This approach carries with it a risk of soil and water pollution through excessive wastewater application. Hence, a detailed understanding of soil water conditions is essential for effective utilization of wastewater for irrigation.

Efficiency

Young engineers restoring and developing the old Mughal irrigation system during the reign of the Mughal Emperor Bahadur Shah II

Modern irrigation methods are efficient enough to supply the entire field uniformly with water, so that each plant has the amount of water it needs, neither too much nor too little. Water use efficiency in the field can be determined as follows:

- Field Water Efficiency (%) = (Water Transpired by Crop ÷ Water Applied to Field) x 100

Until 1960s, the common perception was that water was an infinite resource. At that time, there were fewer than half the current number of people on the planet. People were not as wealthy as today, consumed fewer calories and ate less meat, so less water was needed to produce their food. They required a third of the volume of water we presently take from rivers. Today, the competition for water resources is much more intense. This is because there are now more than seven billion people on the planet, their consumption of water-thirsty meat and vegetables is rising, and there is increasing competition for water from industry, urbanisation and biofuel crops. To avoid a global water crisis, farmers will have to strive to increase productivity to meet growing demands for food, while industry and cities find ways to use water more efficiently.

Successful agriculture is dependent upon farmers having sufficient access to water. However, water scarcity is already a critical constraint to farming in many parts of the world. With regards to agriculture, the World Bank targets food production and water management as an increasingly global issue that is fostering a growing debate. Physical water scarcity is where there is not enough water to meet all demands, including that needed for ecosystems to function effectively. Arid regions frequently suffer from physical water scarcity. It also occurs where water seems abundant but where resources are over-committed. This can happen where there is overdevelopment of hydraulic infrastructure, usually for irrigation. Symptoms of physical water scarcity include environmental degradation and declining groundwater. Economic scarcity, meanwhile, is caused by a lack of investment in water or insufficient human capacity to satisfy the demand for water. Symptoms of economic water scarcity include a lack of infrastructure, with people often having to fetch water from rivers for domestic and agricultural uses. Some 2.8 billion people currently live in water-scarce areas.

Technical Challenges

Irrigation schemes involve solving numerous engineering and economic problems while minimizing negative environmental impact.

- Competition for surface water rights.

- Overdrafting (depletion) of underground aquifers.

- Ground subsidence (e.g. New Orleans, Louisiana)

- Underirrigation or irrigation giving only just enough water for the plant (e.g. in drip line irrigation) gives poor soil salinity control which leads to increased soil salinity with consequent buildup of toxic salts on soil surface in areas with high evaporation. This requires either leaching to remove these salts and a method of drainage to carry the salts away. When using drip lines, the leaching is best done regularly at certain intervals (with only a slight excess of water), so that the salt is flushed back under the plant's roots.

- Overirrigation because of poor distribution uniformity or management wastes water, chemicals, and may lead to water pollution.

- Deep drainage (from over-irrigation) may result in rising water tables which in some instances will lead to problems of irrigation salinity requiring watertable control by some form of subsurface land drainage.

- Irrigation with saline or high-sodium water may damage soil structure owing to the formation of alkaline soil

- Clogging of filters: It is mostly algae that clog filters, drip installations and nozzles. UV and ultrasonic method can be used for algae control in irrigation systems.

Weed Control

Weed control is the botanical component of pest control, which attempts to stop weeds, especially noxious or injurious weeds, from competing with domesticated plants and livestock. Weed control is important in agriculture. Many strategies have been developed in order to contain these plants. Methods include hand cultivation with hoes, powered cultivation with cultivators, smothering with mulch, lethal wilting with high heat, burning, and chemical attack with herbicides (weed killers).

A plant is often termed a "weed" when it has one or more of the following characteristics:

- Little or no recognized value (as in medicinal, material, nutritional or energy)

- Rapid growth and/or ease of germination

- Competitive with crops for space, light, water and nutrients

The definition of a weed is completely context-dependent. To one person, one plant may be a weed, and to another person it may be a desirable plant. In one place, a plant may be viewed as a weed, whereas in another place, the same plant may be desirable.

Introduction

Weeds compete with productive crops or pasture, ultimately converting productive land into unusable scrub. Weeds can be poisonous, distasteful, produce burrs, thorns or otherwise interfere with the use and management of desirable plants by contaminating harvests or interfering with livestock.

Weeds compete with crops for space, nutrients, water and light. Smaller, slower growing seedlings are more susceptible than those that are larger and more vigorous. Onions are one of the most vulnerable, because they are slow to germinate and produce slender, upright stems. By contrast broad beans produce large seedlings and suffer far fewer effects other than during periods of water shortage at the crucial time when the pods are filling out. Transplanted crops raised in sterile soil or potting compost gain a head start over germinating weeds.

Weeds also vary in their competitive abilities and according to conditions and season. Tall-growing vigorous weeds such as fat hen (*Chenopodium album*) can have the most pronounced effects on adjacent crops, although seedlings of fat hen that appear in late summer produce only small plants. Chickweed (*Stellaria media*), a low growing plant, can happily co-exist with a tall crop during the summer, but plants that have overwintered will grow rapidly in early spring and may swamp crops such as onions or spring greens.

The presence of weeds does not necessarily mean that they are damaging a crop, especially during the early growth stages when both weeds and crops can grow without interference. However, as growth proceeds they each begin to require greater amounts of water and nutrients. Estimates suggest that weed and crop can co-exist harmoniously for around three weeks before competition becomes significant. One study found that after competition had started, the final yield of onion bulbs was reduced at almost 4% per day.

Perennial weeds with bulbils, such as lesser celandine and oxalis, or with persistent underground stems such as couch grass (*Agropyron repens*) or creeping buttercup (*Ranunculus repens*) store reserves of food, and are thus able to grow faster and with more vigour than their annual counterparts. Some perennials such as couch grass exude allelopathic chemicals that inhibit the growth of other nearby plants.

Weeds can also host pests and diseases that can spread to cultivated crops. Charlock and Shepherd's purse may carry clubroot, eelworm can be harboured by chickweed, fat hen and shepherd's purse, while the cucumber mosaic virus, which can devastate the cucurbit family, is carried by a range of different weeds including chickweed and groundsel.

Insect pests often do not attack weeds. However pests such as cutworms may first attack weeds then move on to cultivated crops.

Some plants are considered weeds by some farmers and crops by others. Charlock, a common weed in the southeastern US, are weeds according to row crop growers, but are valued by beekeepers, who seek out places where it blooms all winter, thus providing pollen for honeybees and other pollinators. Its bloom resists all but a very hard freeze, and recovers once the freeze ends.

Weed Propagation

Seeds

Annual and biennial weeds such as chickweed, annual meadow grass, shepherd's purse, groundsel, fat hen, cleaver, speedwell and hairy bittercress propagate themselves by seeding. Many produce huge numbers of seed several times a season, some all year round. Groundsel can produce 1000 seed, and can continue right through a mild winter, whilst Scentless Mayweedproduces over 30,000 seeds per plant. Not all of these will germinate at once, but over several seasons, lying dormant in the soil sometimes for years until exposed to light. Poppy seed can survive 80–100 years, dock 50 or more. There can be many thousands of seeds in a square foot or square metre of ground, thus and soil disturbance will produce a flush of fresh weed seedlings.

Subsurface/Surface

The most persistent perennials spread by underground creeping rhizomes that can regrow from a tiny fragment. These include couch grass, bindweed, ground elder, nettles, rosebay willow herb, Japanese knotweed, horsetail and bracken, as well as creeping thistle, whose tap roots can put out lateral roots. Other perennials put out runners that spread along the soil surface. As they creep they set down roots, enabling them to colonise bare ground with great rapidity. These include creeping buttercup and ground ivy. Yet another group of perennials propagate by stolons- stems that arch back into the ground to reroot. The most familiar of these is the bramble.

Methods

Weed control plans typically consist of many methods which are divided into biological, chemical, cultural, and physical/mechanical control.

Pesticide-free thermic weed control with a weed burner on a potato field in Dithmarschen

Physical/Mechanical Methods

Coverings

In domestic gardens, methods of weed control include covering an area of ground with a material that creates a hostile environment for weed growth, known as a *weed mat*.

Several layers of wet newspaper prevent light from reaching plants beneath, which kills them. Daily saturating the newspaper with water plant decomposition. After several weeks, all germinating weed seeds are dead.

In the case of black plastic, the greenhouse effect kills the plants. Although the black plastic sheet is effective at preventing weeds that it covers, it is difficult to achieve complete coverage. Eradicating persistent perennials may require the sheets to be left in place for at least two seasons.

Some plants are said to produce root exudates that suppress herbaceous weeds. *Tagetes minuta* is claimed to be effective against couch and ground elder, whilst a border of comfrey is also said to act as a barrier against the invasion of some weeds including couch. A 5–10 centimetres (2.0–3.9 in)} layer of wood chip mulch prevents most weeds from sprouting.

Gravel can serve as an inorganic mulch.

Irrigation is sometimes used as a weed control measure such as in the case of paddy fields to kill any plant other than the water-tolerant rice crop.

Manual Removal

Weeds are removed manually in large parts of India.

Many gardeners still remove weeds by manually pulling them out of the ground, making sure to include the roots that would otherwise allow them to resprout.

Hoeing off weed leaves and stems as soon as they appear can eventually weaken and kill perennials, although this will require persistence in the case of plants such as bindweed. Nettle infestations can be tackled by cutting back at least three times a year, repeated over a three-year period. Bramble can be dealt with in a similar way.

Tillage

Ploughing includes tilling of soil, intercultural ploughing and summer ploughing. Ploughing uproots weeds, causing them to die. In summer ploughing is done during deep summers. Summer ploughing also helps in killing pests.

Mechanical tilling can remove weeds around crop plants at various points in the growing process.

Thermal

Several thermal methods can control weeds.

Flame weeders use a flame several centimeters away from the weeds to give them a sudden and severe heating. The goal of flame weeding is not necessarily burning the plant, but rather causing a lethal wilting by denaturing proteins in the weed. Similarly, hot air weeders can heat up the seeds to the point of destroying them. Flame weeders can be combined with techniques such as stale seedbeds (preparing and watering the seedbed early, then killing the nascent crop of weeds that springs up from it, then sowing the crop seeds) and preemergence flaming (doing a flame pass against weed seedlings after the sowing of the crop seeds but before those seedlings emerge from the soil—a span of time that can be days or weeks).

Hot foam (foamstream) causes the cell walls to rupture, killing the plant. Weed burners heat up soil quickly and destroy superficial parts of the plants. Weed seeds are often heat resistant and even react with an increase of growth on dry heat.

Since the 19th century soil steam sterilization has been used to clean weeds completely from soil. Several research results confirm the high effectiveness of humid heat against weeds and its seeds.

Soil solarization in some circumstances is very effective at eliminating weeds while maintaining grass. Planted grass tends to have a higher heat/humidity tolerance than unwanted weeds.

Seed Targeting

In 1998, the Australian Herbicide Resistance Initiative (AHRI), debuted. gathered fifteen scientists and technical staff members to conduct field surveys, collect seeds, test for resistance and study the biochemical and genetic mechanisms of resistance. A collaboration with DuPont led to a mandatory herbicide labeling program, in which each mode of action is clearly identified by a letter of the alphabet.

The key innovation of the AHRI approach has been to focus on weed seeds. Ryegrass seeds last only a few years in soil, so if farmers can prevent new seeds from arriving, the number of sprouts will shrink each year. Until the new approach farmers were unintentionally helping the seeds. Their combines loosen ryegrass seeds from their stalks and spread them over the fields. In the mid-1980s, a few farmers hitched covered trailers, called "chaff carts", behind their combines to catch the chaff and weed seeds. The collected material is then burned.

An alternative is to concentrate the seeds into a half-meter-wide strip called a windrow and burn the windrows after the harvest, destroying the seeds. Since 2003, windrow burning has been adopted by about 70% of farmers in Western Australia.

Yet another approach is the Harrington Seed Destructor, which is an adaptation of a coal pulverizing cage mill that uses steel bars whirling at up to 1500 rpm. It keeps all the organic material in the field and does not involve combustion, but kills 95% of seeds.

Cultural Methods

Stale Seed Bed

Another manual technique is the 'stale seed bed', which involves cultivating the soil, then leaving it fallow for a week or so. When the initial weeds sprout, the grower lightly hoes them away before planting the desired crop. However, even a freshly cleared bed is susceptible to airborne seed from elsewhere, as well as seed carried by passing animals on their fur, or from imported manure.

Buried Drip Irrigation

Buried drip irrigation involves burying drip tape in the subsurface near the planting bed, thereby limiting weeds access to water while also allowing crops to obtain moisture. It is most effective during dry periods.

Crop Rotation

Rotating crops with ones that kill weeds by choking them out, such as hemp, *Mucuna pruriens*, and other crops, can be a very effective method of weed control. It is a way to avoid the use of herbicides, and to gain the benefits of crop rotation.

Biological Methods

A biological weed control regiment can consist of biological control agents, bioherbicides, use of grazing animals, and protection of natural predators.

Animal Grazing

Companies using goats to control and eradicate leafy spurge, knapweed, and other toxic weeds have sprouted across the American West.

Chemical Methods

"Organic" Approaches

Weed control, circa 1930-40s

A mechanical weed control device: the diagonal weeder

Organic weed control involves anything other than applying manufactured chemicals. Typically a combination of methods are used to achieve satisfactory control.

Sulfur in some circumstances is accepted within British Soil Association standards.

Herbicides

The above described methods of weed control use no or very limited chemical inputs. They are preferred by organic gardeners or organic farmers.

However weed control can also be achieved by the use of herbicides. Selective herbicides kill certain targets while leaving the desired crop relatively unharmed. Some of these act by interfering with the growth of the weed and are often based on plant hormones. Herbicides are generally classified as follows:

- Contact herbicides destroy only plant tissue that contacts the herbicide. Generally, these are the fastest-acting herbicides. They are ineffective on perennial plants that can re-grow from roots or tubers.

- Systemic herbicides are foliar-applied and move through the plant where they destroy a greater amount of tissue. Glyphosate is currently the most used systemic herbicide.

- Soil-borne herbicides are applied to the soil and are taken up by the roots of the target plant.

- Pre-emergent herbicides are applied to the soil and prevent germination or early growth of weed seeds.

In agriculture large scale and systematic procedures are usually required, often by machines, such as large liquid herbicide 'floater' sprayers, or aerial application.

Bradley Method

Bradley Method of Bush Regeneration, which uses ecological processes to do much of the work. Perennial weeds also propagate by seeding; the airborne seed of the dandelion and the rose-bay willow herb parachute far and wide. Dandelion and dock also put down deep tap roots, which, although they do not spread underground, are able to regrow from any remaining piece left in the ground.

Hybrid

One method of maintaining the effectiveness of individual strategies is to combine them with others that work in complete different ways. Thus seed targeting has been combined with herbicides. In Australia seed management has been effectively combined with trifluralin and clethodim.

Resistance

Resistance occurs when a target adapts to circumvent a particular control strategy. It affects not only weed control,but antibiotics, insect control and other domains. In agriculture is mostly considered in reference to pesticides, but can defeat other strategies, e.g., when a target species becomes more drought tolerant via selection pressure.

Farming Practices

Herbicide resistance recently became a critical problem as many Australian sheep farmers switched to exclusively growing wheat in their pastures in the 1970s. In wheat fields, introduced varieties of ryegrass, while good for grazing sheep, are intense competitors with wheat. Ryegrasses produce so many seeds that, if left unchecked, they can completely choke a field. Herbicides provided excellent control, while reducing soil disrupting because of less need to plough. Within little more than a decade, ryegrass and other weeds began to develop resistance. Australian farmers evolved again and began diversifying their techniques.

In 1983, patches of ryegrass had become immune to Hoegrass, a family of herbicides that inhibit an enzyme called acetyl coenzyme A carboxylase.

Ryegrass populations were large, and had substantial genetic diversity, because farmers had planted many varieties. Ryegrass is cross-pollinated by wind, so genes shuffle frequently. Farmers sprayed inexpensive Hoegrass year after year, creating selection pressure, but were diluting the herbicide in order to save money, increasing plants survival. Hoegrass was mostly replaced by a group of herbicides that block acetolactate synthase, again helped by poor application practices. Ryegrass evolved a kind of "cross-resistance" that allowed it to rapidly break down a variety of herbicides. Australian farmers lost four classes of herbicides in only a few years. As of 2013 only two herbicide classes, called Photosystem II and long-chain fatty acid inhibitors, had become the last hope.

Terrace (Agriculture)

Rice terrace in Indonesia

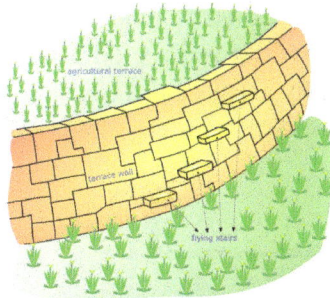

Diagram showing Inca terrace engineering for agriculture.

In agriculture, a terrace is a piece of sloped plane that has been cut into a series of successively receding flat surfaces or platforms, which resemble steps, for the purposes of more effective farming . This type of landscaping, therefore, is called terracing. Graduated terrace steps are commonly used to farm on hilly or mountainous terrain. Terraced fields decrease both erosion and surface runoff, and may be used to support growing crops that require irrigation, such as rice. The Rice Terraces of the Philippine Cordilleras have been designated as a UNESCO World Heritage Site because of the significance of this technique.

Terraced paddy fields are used widely in rice, wheat and barley farming in east, south, and southeast Asia, as well as other places. Drier-climate terrace farming is common throughout the Mediterranean Basin, e.g., in Cadaqués, Catalonia, where they were used for vineyards, olive trees, cork oak, etc., on Mallorca, and in Cinque Terre, Italy.

In the South American Andes, farmers have used terraces, known as *andenes*, for over a thousand years to farm potatoes, maize, and other native crops. Terraced farming was developed by the Wari' and other peoples of the south-central Andes before 1000 AD, centuries before they were used by the Inca, who adopted them. The terraces were built to make the most efficient use of shallow soil and to enable irrigation of crops.

The Inca built on these, developing a system of canals, aqueducts, and puquios to direct water through dry land and increase fertility levels and growth. These terraced farms are found wherever mountain villages have existed in the Andes. They provided the food necessary to support the populations of great Inca cities and religious centres such as Machu Picchu.

Terracing is also used for sloping terrain; the Hanging Gardens of Babylon may have been built on an artificial mountain with stepped terraces, such as those on a ziggurat. At the seaside Villa of the Papyri in Herculaneum, the villa gardens of Julius Caesar's father-in-law were designed in terraces to give pleasant and varied views of the Bay of Naples.

Terraced fields are common in islands with steep slopes. The Canary Islands present a complex system of terraces covering the landscape from the coastal irrigated plantations to the dry fields in the highlands. These terraces, which are named *cadenas* (chains), are built with stone walls of skillful design, which include attached stairs and channels.

In Old English, a terrace was also called a "lynch" (lynchet). An example of an ancient Lynch Mill is in Lyme Regis. The water is directed from a river by a duct along a terrace. This set-up was used in steep hilly areas in the UK.

Aquaculture

Aquaculture installations in southern Chile

Aquaculture, also known as aquafarming, is the farming of fish, crustaceans, molluscs, aquatic plants, algae, and other aquatic organisms. Aquaculture involves cultivating freshwater and saltwater populations under controlled conditions, and can be contrasted with commercial fishing, which is the harvesting of wild fish. Mariculture refers to aquaculture practiced in marine environments and in underwater habitats.

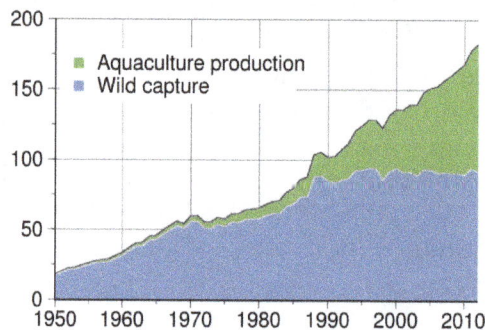

Global harvest of aquatic organisms in million tonnes, 1950–2010, as reported by the FAO

According to the FAO, aquaculture "Farming implies some form of intervention in the rearing process to enhance production, such as regular stocking, feeding, protection from predators, etc. Farming also implies individual or corporate ownership of the stock being cultivated." The reported output from global aquaculture operations in 2014 supplied over one half of the fish and shellfish that is directly consumed by humans; however, there are issues about the reliability of the reported figures. Further, in current aquaculture practice, products from several pounds of wild fish are used to produce one pound of a piscivorous fish like salmon.

Particular kinds of aquaculture include fish farming, shrimp farming, oyster farming, mariculture, algaculture (such as seaweed farming), and the cultivation of ornamental fish. Particular methods include aquaponics and integrated multi-trophic aquaculture, both of which integrate fish farming and plant farming.

History

Workers harvest catfish from the Delta Pride Catfish farms in Mississippi

The indigenous Gunditjmara people in Victoria, Australia, may have raised eels as early as 6000 BC. Evidence indicates they developed about 100 km² (39 sq mi) of volcanic floodplains in the vicinity of Lake Condah into a complex of channels and dams, and used woven traps to capture eels, and preserve them to eat all year round.

Aquaculture was operating in China *circa* 2500 BC. When the waters subsided after river floods, some fish, mainly carp, were trapped in lakes. Early aquaculturists fed their brood using nymphs and silkworm feces, and ate them. A fortunate genetic mutation of carp led to the emergence of goldfish during the Tang dynasty.

Japanese cultivated seaweed by providing bamboo poles and, later, nets and oyster shells to serve as anchoring surfaces for spores.

Romans bred fish in ponds and farmed oysters in coastal lagoons before 100 CE.

In central Europe, early Christian monasteries adopted Roman aquacultural practices. Aquaculture spread in Europe during the Middle Ages since away from the seacoasts and the big rivers, fish had to be salted so they did not rot. Improvements in transportation during the 19th century made fresh fish easily available and inexpensive, even in inland areas, making aquaculture less popular. The 15th-century fishponds of the Trebon Basin in the Czech Republic are maintained as a UNESCO World Heritage Site.

Hawaiians constructed oceanic fish ponds. A remarkable example is a fish pond dating from at least 1,000 years ago, at Alekoko. Legend says that it was constructed by the mythical Menehune dwarf people.

In first half of 18th century, German Stephan Ludwig Jacobi experimented with external fertilization of brown trouts and salmon. He wrote an article *"Von der künstlichen Erzeugung der Forellen und Lachse"*. By the latter decades of the 18th century, oyster farming had begun in estuaries along the Atlantic Coast of North America.

The word aquaculture appeared in an 1855 newspaper article in reference to the harvesting of ice. It also appeared in descriptions of the terrestrial agricultural practise of subirrigation in the late 19th century before becoming associated primarily with the cultivation of aquatic plant and animal species.

In 1859, Stephen Ainsworth of West Bloomfield, New York, began experiments with brook trout. By 1864, Seth Green had established a commercial fish-hatching operation at Caledonia Springs, near Rochester, New York. By 1866, with the involvement of Dr. W. W. Fletcher of Concord, Massachusetts, artificial fish hatcheries were under way in both Canada and the United States. When the Dildo Island fish hatchery opened in Newfoundland in 1889, it was the largest and most advanced in the world. The word aquaculture was used in descriptions of the hatcheries experiments with cod and lobster in 1890.

By the 1920s, the American Fish Culture Company of Carolina, Rhode Island, founded in the 1870s was one of the leading producers of trout. During the 1940s, they had perfected the method of manipulating the day and night cycle of fish so that they could be artificially spawned year around.

Californians harvested wild kelp and attempted to manage supply around 1900, later labeling it a wartime resource.

21st-century Practice

Harvest stagnation in wild fisheries and overexploitation of popular marine species, combined with a growing demand for high-quality protein, encouraged aquaculturists to domesticate other marine species. At the outset of modern aquaculture, many were optimistic that a "Blue Revolution" could take place in aquaculture, just as the Green Revolution of the 20th century had revolutionized agriculture. Although land animals had long been domesticated, most seafood species were still caught from the wild. Concerned about the impact of growing demand for seafood on the world's oceans, prominent ocean explorer Jacques Cousteau wrote in 1973: "With earth's burgeoning human populations to feed, we must turn to the sea with new understanding and new technology."

About 430 (97%) of the species cultured as of 2007 were domesticated during the 20th and 21st centuries, of which an estimated 106 came in the decade to 2007. Given the long-term importance of agriculture, to date, only 0.08% of known land plant species and 0.0002% of known land animal species have been domesticated, compared with 0.17% of known marine plant species and 0.13% of known marine animal species. Domestication typically involves about a decade of scientific research. Domesticating aquatic species involves fewer risks to humans than do land animals, which took a large toll in human lives. Most major human diseases originated in domesticated animals, including diseases such as smallpox and diphtheria, that like most infectious diseases, move to humans from animals. No human pathogens of comparable virulence have yet emerged from marine species.

Biological control methods to manage parasites are already being used, such as cleaner fish (e.g. lumpsuckers and wrasse) to control sea lice populations in salmon farming. Models are being used to help with spatial planning and siting of fish farms in order to minimize impact.

The decline in wild fish stocks has increased the demand for farmed fish. However, finding alternative sources of protein and oil for fish feed is necessary so the aquaculture industry can grow sustainably; otherwise, it represents a great risk for the over-exploitation of forage fish.

Another recent issue following the banning in 2008 of organotins by the International Maritime Organization is the need to find environmentally friendly, but still effective, compounds with antifouling effects.

Many new natural compounds are discovered every year, but producing them on a large enough scale for commercial purposes is almost impossible.

It is highly probable that future developments in this field will rely on microorganisms, but greater funding and further research is needed to overcome the lack of knowledge in this field.

Species Groups

Global Aquaculture Production in Million Tonnes, 1950–2010, as Reported by the FAO

Main species groups

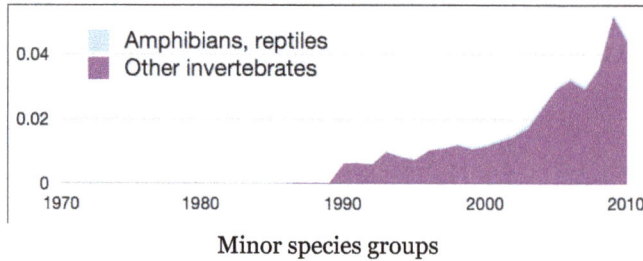

Minor species groups

Aquatic Plants

Cultivating emergent aquatic plants in floating containers

Microalgae, also referred to as phytoplankton, microphytes, or planktonic algae, constitute the majority of cultivated algae. Macroalgae commonly known as seaweed also have many commercial and industrial uses, but due to their size and specific requirements, they are not easily cultivated on a large scale and are most often taken in the wild.

Fish

The farming of fish is the most common form of aquaculture. It involves raising fish commercially in tanks, ponds, or ocean enclosures, usually for food. A facility that releases juvenile fish into the wild for recreational fishing or to supplement a species' natural numbers is generally referred to as a fish hatchery. Worldwide, the most important fish species used in fish farming are, in order, carp, salmon, tilapia, and catfish.

In the Mediterranean, young bluefin tuna are netted at sea and towed slowly towards the shore. They are then interned in offshore pens where they are further grown for the market. In 2009, researchers in Australia managed for the first time to coax southern bluefin tuna to breed in landlocked tanks. Southern bluefin tuna are also caught in the wild and fattened in grow-out sea cages in southern Spencer Gulf, South Australia.

A similar process is used in the salmon-farming section of this industry; juveniles are taken from hatcheries and a variety of methods are used to aid them in their maturation. For example, as stated above, some of the most important fish species in the industry, salmon, can be grown using a cage system. This is done by having netted cages, preferably in open water that has a strong flow, and feeding the salmon a special food mixture that aids their growth. This process allows for year-round growth of the fish, thus a higher harvest during the correct seasons. An additional method, known sometimes as sea ranching, has also been used within the industry. Sea ranching involves raising fish in a hatchery for a brief time and then releasing them into marine waters for further development, whereupon the fish are recaptured when they have matured.

Crustaceans

Commercial shrimp farming began in the 1970s, and production grew steeply thereafter. Global production reached more than 1.6 million tonnes in 2003, worth about US$9 billion. About 75% of farmed shrimp is produced in Asia, in particular in China and Thailand. The other 25% is produced mainly in Latin America, where Brazil is the largest producer. Thailand is the largest exporter.

Shrimp farming has changed from its traditional, small-scale form in Southeast Asia into a global industry. Technological advances have led to ever higher densities per unit area, and broodstock is shipped worldwide. Virtually all farmed shrimp are penaeids (i.e., shrimp of the family Penaeidae), and just two species of shrimp, the Pacific white shrimp and the giant tiger prawn, account for about 80% of all farmed shrimp. These industrial monocultures are very susceptible to disease, which has decimated shrimp populations across entire regions. Increasing ecological problems, repeated disease outbreaks, and pressure and criticism from both nongovernmental organizations and consumer countries led to changes in the industry in the late 1990s and generally stronger regulations. In 1999, governments, industry representatives, and environmental organizations initiated a program aimed at developing and promoting more sustainable farming practices through the Seafood Watch program.

Freshwater prawn farming shares many characteristics with, including many problems with, marine shrimp farming. Unique problems are introduced by the developmental lifecycle of the main species, the giant river prawn.

The global annual production of freshwater prawns (excluding crayfish and crabs) in 2003 was about 280,000 tonnes, of which China produced 180,000 tonnes followed by India and Thailand with 35,000 tonnes each. Additionally, China produced about 370,000 tonnes of Chinese river crab.

Molluscs

Aquacultured shellfish include various oyster, mussel, and clam species. These bivalves are filter and/or deposit feeders, which rely on ambient primary production rather than inputs of fish or other feed. As such, shellfish aquaculture is generally perceived as benign or even beneficial.

Abalone farm

Depending on the species and local conditions, bivalve molluscs are either grown on the beach, on longlines, or suspended from rafts and harvested by hand or by dredging.

Abalone farming began in the late 1950s and early 1960s in Japan and China. Since the mid-1990s, this industry has become increasingly successful. Overfishing and poaching have reduced wild populations to the extent that farmed abalone now supplies most abalone meat. Sustainably farmed molluscs can be certified by Seafood Watch and other organizations, including the World Wildlife Fund (WWF). WWF initiated the "Aquaculture Dialogues" in 2004 to develop measurable and performance-based standards for responsibly farmed seafood. In 2009, WWF co-founded the Aquaculture Stewardship Council with the Dutch Sustainable Trade Initiative to manage the global standards and certification programs.

After trials in 2012, a commercial "sea ranch" was set up in Flinders Bay, Western Australia, to raise abalone. The ranch is based on an artificial reef made up of 5000 (As of April 2016) separate concrete units called 'abitats' (abalone habitats). The 900-kg abitats can host 400 abalone each. The reef is seeded with young abalone from an onshore hatchery. The abalone feed on seaweed that has grown naturally on the abitats, with the ecosystem enrichment of the bay also resulting in growing numbers of dhufish, pink snapper, wrasse, and Samson fish, among other species.

Brad Adams, from the company, has emphasised the similarity to wild abalone and the difference from shore-based aquaculture. "We're not aquaculture, we're ranching, because once they're in the water they look after themselves."

Other Groups

Other groups include aquatic reptiles, amphibians, and miscellaneous invertebrates, such as echinoderms and jellyfish. They are separately graphed at the top right of this section, since they do not contribute enough volume to show clearly on the main graph.

Commercially harvested echinoderms include sea cucumbers and sea urchins. In China, sea cucumbers are farmed in artificial ponds as large as 1,000 acres (400 ha).

Around the World

Global Aquaculture Production in Million Tonnes, 1950–2010, as Reported by the FAO

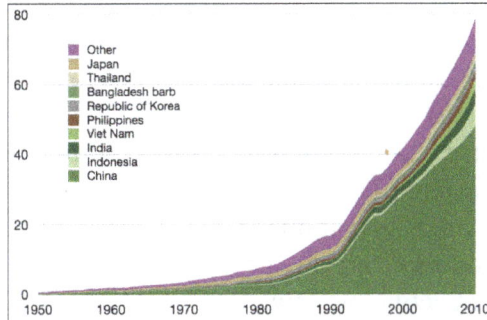

Main aquaculture countries, 1950–2010

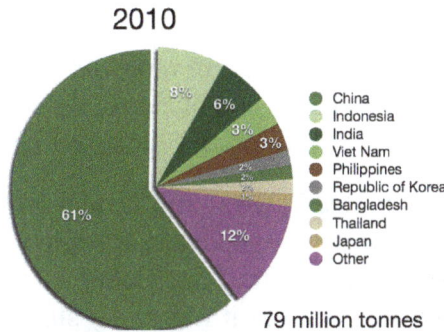

Main aquaculture countries in 2010

In 2012, the total world production of fisheries was 158 million tonnes, of which aquaculture contributed 66.6 million tonnes, about 42%. The growth rate of worldwide aquaculture has been sustained and rapid, averaging about 8% per year for over 30 years, while the take from wild fisheries] has been essentially flat for the last decade. The aquaculture market reached $86 billion in 2009.

Aquaculture is an especially important economic activity in China. Between 1980 and 1997, the Chinese Bureau of Fisheries reports, aquaculture harvests grew at an annual rate of 16.7%, jumping from 1.9 million tonnes to nearly 23 million tonnes. In 2005, China accounted for 70% of world production. Aquaculture is also currently one of the fastest-growing areas of food production in the U.S.

About 90% of all U.S. shrimp consumption is farmed and imported. In recent years, salmon aquaculture has become a major export in southern Chile, especially in Puerto Montt, Chile's fastest-growing city.

A United Nations report titled *The State of the World Fisheries and Aquaculture* released in May 2014 maintained fisheries and aquaculture support the livelihoods of some 60 million people in Asia and Africa.

National Laws, Regulations, and Management

Laws governing aquaculture practices vary greatly by country and are often not closely regulated or easily traceable. In the United States, land-based and nearshore aquaculture is regulated at the federal and state levels; however, no national laws govern offshore aquaculture in U.S. exclusive economic zone waters. In June 2011, the Department of Commerce and National Oceanic and Atmospheric Administration released national aquaculture policies to address this issue and "to meet the growing demand for healthy seafood, to create jobs in coastal communities, and restore vital ecosystems." In 2011, Congresswoman Lois Capps introduced the *National Sustainable Offshore Aquaculture Act of 2011* "to establish a regulatory system and research program for sustainable offshore aquaculture in the United States exclusive economic zone"; however, the bill was not enacted into law.

Over-reporting

China overwhelmingly dominates the world in reported aquaculture output, reporting a total output which is double that of the rest of the world put together. However, there are some historical issues with the accuracy of China's returns.

In 2001, the fisheries scientists Reg Watson and Daniel Pauly expressed concerns in a letter to *Nature*, that China was over reporting its catch from wild fisheries in the 1990s.. They said that made it appear that the global catch since 1988 was increasing annually by 300,000 tonnes, whereas it was really shrinking annually by 350,000 tonnes. Watson and Pauly suggested this may be have been related to Chinese policies where state entities that monitored the economy were also tasked with increasing output. Also, until more recently, the promotion of Chinese officials was based on production increases from their own areas.

China disputed this claim. The official Xinhua News Agency quoted Yang Jian, director general of the Agriculture Ministry's Bureau of Fisheries, as saying that China's figures were "basically correct". However, the FAO accepted there were issues with the reliability of China's statistical returns, and for a period treated data from China, including the aquaculture data, apart from the rest of the world.

Aquacultural Methods

Mariculture

Mariculture refers to the cultivation of marine organisms in seawater, usually in sheltered coastal waters. The farming of marine fish is an example of mariculture, and so also is the farming of marine crustaceans (such as shrimp), molluscs (such as oysters), and seaweed.

Mariculture may consist of raising the organisms on or in artificial enclosures such as in floating netted enclosures for salmon and on racks for oysters. In the case of enclosed salmon, they are fed by the operators; oysters on racks filter feed on naturally available food. Abalone have been farmed on an artificial reef consuming seaweed which grows naturally on the reef units.

Mariculture off High Island, Hong Kong

The adaptable tilapia is another commonly farmed fish

Integrated

Integrated multi-trophic aquaculture (IMTA) is a practice in which the byproducts (wastes) from one species are recycled to become inputs (fertilizers, food) for another. Fed aquaculture (for example, fish, shrimp) is combined with inorganic extractive and organic extractive (for example, shellfish) aquaculture to create balanced systems for environmental sustainability (biomitigation), economic stability (product diversification and risk reduction) and social acceptability (better management practices).

"Multi-trophic" refers to the incorporation of species from different trophic or nutritional levels in the same system. This is one potential distinction from the age-old practice of aquatic polyculture, which could simply be the co-culture of different fish species from the same trophic level. In this case, these organisms may all share the same biological and chemical processes, with few synergistic benefits, which could potentially lead to significant shifts in the ecosystem. Some traditional polyculture systems may, in fact, incorporate a greater diversity of species, occupying several niches, as extensive cultures (low intensity, low management) within the same pond. The term "integrated" refers to the more intensive cultivation of the different species in proximity of each other, connected by nutrient and energy transfer through water.

Ideally, the biological and chemical processes in an IMTA system should balance. This is achieved through the appropriate selection and proportions of different species providing different ecosystem functions. The co-cultured species are typically more than just biofilters; they are harvestable crops of commercial value. A working IMTA system can result in greater total production based on mutual benefits to the co-cultured species and improved ecosystem health, even if the production of individual species is lower than in a monoculture over a short term period.

Sometimes the term "integrated aquaculture" is used to describe the integration of monocultures through water transfer. For all intents and purposes, however, the terms "IMTA" and "integrated aquaculture" differ only in their degree of descriptiveness. Aquaponics, fractionated aquaculture, integrated agriculture-aquaculture systems, integrated peri-urban-aquaculture systems, and integrated fisheries-aquaculture systems are other variations of the IMTA concept.

Netting Materials

Various materials, including nylon, polyester, polypropylene, polyethylene, plastic-coated welded wire, rubber, patented rope products (Spectra, Thorn-D, Dyneema), galvanized steel and copper are used for netting in aquaculture fish enclosures around the world. All of these materials are selected for a variety of reasons, including design feasibility, material strength, cost, and corrosion resistance.

Recently, copper alloys have become important netting materials in aquaculture because they are antimicrobial (i.e., they destroy bacteria, viruses, fungi, algae, and other microbes) and they therefore prevent biofouling (i.e., the undesirable accumulation, adhesion, and growth of microorganisms, plants, algae, tubeworms, barnacles, mollusks, and other organisms). By inhibiting microbial growth, copper alloy aquaculture cages avoid costly net changes that are necessary with other materials. The resistance of organism growth on copper alloy nets also provides a cleaner and healthier environment for farmed fish to grow and thrive.

Issues

If performed without consideration for potential local environmental impacts, aquaculture in inland waters can result in more environmental damaging than wild fisheries, though with less waste produced on a per kg on a global scale. Local concerns with aquaculture in inland waters may include waste handling, side-effects of antibiotics, competition between farmed and wild animals, and the potential introduction of invasive plant and animal species, or foreign pathogens, particularly if unprocessed fish are used to feed more marketable carnivorous fish. If non-local live feeds are used, aquaculture may introduce plant of animal. Improvements in methods resulting from advances in research and the availability of commercial feeds has reduced some of these concerns since their greater prevalence in 1990s and 2000s .

Fish waste is organic and composed of nutrients necessary in all components of aquatic food webs. In-ocean aquaculture often produces much higher than normal fish waste concentrations. The waste collects on the ocean bottom, damaging or eliminating bottom-dwelling life. Waste can also decrease dissolved oxygen levels in the water column, putting further pressure on wild animals. An alternative model to food being added to the ecosystem, is the installation of artificial reef structures to increase the habitat niches available, without the need to add any more than ambient feed and nutrient. This has been used in the "ranching" of abalone in Western Australia.

Fish Oils

Tilapia from aquaculture has been shown to contain more fat and a much higher ratio of omega-6 to omega-3 oils.

Impacts on Wild Fish

Some carnivorous and omnivorous farmed fish species are fed wild forage fish. Although carnivorous farmed fish represented only 13 percent of aquaculture production by weight in 2000, they represented 34 percent of aquaculture production by value.

Farming of carnivorous species like salmon and shrimp leads to a high demand for forage fish to match the nutrition they get in the wild. Fish do not actually produce omega-3 fatty acids, but instead accumulate them from either consuming microalgae that produce these fatty acids, as is the case with forage fish like herring and sardines, or, as is the case with fatty predatory fish, like salmon, by eating prey fish that have accumulated omega-3 fatty acids from microalgae. To satisfy this requirement, more than 50 percent of the world fish oil production is fed to farmed salmon.

Farmed salmon consume more wild fish than they generate as a final product, although the efficiency of production is improving. To produce one pound of farmed salmon, products from several pounds of wild fish are fed to them - this can be described as the "fish-in-fish-out" (FIFO) ratio. In 1995, salmon had a FIFO ratio of 7.5 (meaning 7.5 pounds of wild fish feed were required to produce 1 pound of salmon); by 2006 the ratio had fallen to 4.9. Additionally, a growing share of fish oil and fishmeal come from residues (byproducts of fish processing), rather than dedicated whole fish. In 2012, 34 percent of fish oil and 28 percent of fishmeal came from residues. However, fishmeal and oil from residues instead of whole fish have a different composition with more ash and less protein, which may limit its potential use for aquaculture.

As the salmon farming industry expands, it requires more wild forage fish for feed, at a time when seventy five percent of the worlds monitored fisheries are already near to or have exceeded their maximum sustainable yield. The industrial scale extraction of wild forage fish for salmon farming then impacts the survivability of the wild predator fish who rely on them for food. An important step in reducing the impact of aquaculture on wild fish is shifting carnivorous species to plant-based feeds. Salmon feeds, for example, have gone from containing only fishmeal and oil to containing 40 percent plant protein. The USDA has also experimented with using grain-based feeds for farmed trout. When properly formulated (and often mixed with fishmeal or oil), plant-based feeds can provide proper nutrition and similar growth rates in carnivorous farmed fish.

Another impact aquaculture production can have on wild fish is the risk of fish escaping from coastal pens, where they can interbreed with their wild counterparts, diluting wild genetic stocks. Escaped fish can become invasive, out-competing native species.

Coastal Ecosystems

Aquaculture is becoming a significant threat to coastal ecosystems. About 20 percent of mangrove forests have been destroyed since 1980, partly due to shrimp farming. An extended cost–benefit analysis of the total economic value of shrimp aquaculture built on mangrove ecosystems found that the external costs were much higher than the external benefits. Over four decades, 269,000 hectares (660,000 acres) of Indonesian mangroves have been converted to shrimp farms. Most of these farms are abandoned within a decade because of the toxin build-up and nutrient loss.

Pollution from Sea Cage Aquaculture

Salmon farms are typically sited in pristine coastal ecosystems which they then pollute. A farm with 200,000 salmon discharges more fecal waste than a city of 60,000 people. This waste is discharged directly into the surrounding aquatic environment, untreated, often containing antibiotics and pesticides." There is also an accumulation of heavy metals on the benthos (seafloor) near the salmon farms, particularly copper and zinc.

In 2016, mass fish kill events impacted salmon farmers along Chile's coast and the wider ecology. Increases in aquaculture production and its associated effluent were considered to be possible contributing factors to fish and molluscan mortality.

Sea cage aquaculture is responsible for nutrient enrichment of the waters in which they are established. This results from fish wastes and uneaten feed inputs. Elements of most concern are nitrogen and phosphorus which can promote algal growth, including harmful algal blooms which can be toxic to fish. Flushing times, current speeds, distance from the shore and water depth are important considerations when locating sea cages in order to minimize the impacts of nutrient enrichment on coastal ecosystems.

The extent of the effects of pollution from sea-cage aquaculture varies depending on where the cages are located, which species are kept, how densely cages are stocked and what the fish are fed. Important species-specific variables include the species' food conversion ratio (FCR) and nitrogen retention. Studies prior to 2001 determined that the amount of nitrogen introduced as feed which is lost to the water column and seafloor as waste varies from 52 to 95%.

Genetic Modification

A type of salmon called the AquAdvantage salmon has been genetically modified for faster growth, although it has not been approved for commercial use, due to controversy. The altered salmon incorporates a growth hormone from a Chinook salmon that allows it to reach full size in 16-28 months, instead of the normal 36 months for Atlantic salmon, and while consuming 25 percent less feed. The U.S. Food and Drug Administration reviewed the AquAdvantage salmon in a draft environmental assessment and determined that it "would not have a significant impact (FONSI) on the U.S. environment."

Animal Welfare

As with the farming of terrestrial animals, social attitudes influence the need for humane practices and regulations in farmed marine animals. Under the guidelines advised by the Farm Animal Welfare Council good animal welfare means both fitness and a sense of well being in the animal's physical and mental state. This can be defined by the Five Freedoms:

- Freedom from hunger & thirst

- Freedom from discomfort

- Freedom from pain, disease, or injury

- Freedom to express normal behaviour

- Freedom from fear and distress

However, the controversial issue in aquaculture is whether fish and farmed marine invertebrates are actually sentient, or have the perception and awareness to experience suffering. Although no evidence of this has been found in marine invertebrates, recent studies conclude that fish do have the necessary receptors (nociceptors) to sense noxious stimuli and so are likely to experience states of pain, fear and stress. Consequently, welfare in aquaculture is directed at vertebrates; finfish in particular.

Common Welfare Concerns

Welfare in aquaculture can be impacted by a number of issues such as stocking densities, behavioural interactions, disease and parasitism. A major problem in determining the cause of impaired welfare is that these issues are often all interrelated and influence each other at different times.

Optimal stocking density is often defined by the carrying capacity of the stocked environment and the amount of individual space needed by the fish, which is very species specific. Although behavioural interactions such as shoaling may mean that high stocking densities are beneficial to some species, in many cultured species high stocking densities may be of concern. Crowding can constrain normal swimming behaviour, as well as increase aggressive and competitive behaviours such as cannibalism, feed competition, territoriality and dominance/subordination hierarchies. This potentially increases the risk of tissue damage due to abrasion from fish-to-fish contact or fish-to-cage contact. Fish can suffer reductions in food intake and food conversion efficiency. In addition, high stocking densities can result in water flow being insufficient, creating inadequate oxygen supply and waste product removal. Dissolved oxygen is essential for fish respiration and concentrations below critical levels can induce stress and even lead to asphyxiation. Ammonia, a nitrogen excretion product, is highly toxic to fish at accumulated levels, particularly when oxygen concentrations are low.

Many of these interactions and effects cause stress in the fish, which can be a major factor in facilitating fish disease. For many parasites, infestation depends on the host's degree of mobility, the density of the host population and vulnerability of the host's defence system. Sea lice are the primary parasitic problem for finfish in aquaculture, high numbers causing widespread skin erosion and haemorrhaging, gill congestion, and increased mucus production. There are also a number of prominent viral and bacterial pathogens that can have severe effects on internal organs and nervous systems.

Improving Welfare

The key to improving welfare of marine cultured organisms is to reduce stress to a minimum, as prolonged or repeated stress can cause a range of adverse effects. Attempts to minimise stress can occur throughout the culture process. During grow out it is important to keep stocking densities at appropriate levels specific to each species, as well as separating size classes and grading to reduce aggressive behavioural interactions. Keeping nets and cages clean can assist positive water flow to reduce the risk of water degradation.

Not surprisingly disease and parasitism can have a major effect on fish welfare and it is important for farmers not only to manage infected stock but also to apply disease prevention measures. However, prevention methods, such as vaccination, can also induce stress because of the extra handling and injection. Other methods include adding antibiotics to feed, adding chemicals into water for treatment baths and biological control, such as using cleaner wrasse to remove lice from farmed salmon.

Many steps are involved in transport, including capture, food deprivation to reduce faecal contamination of transport water, transfer to transport vehicle via nets or pumps, plus transport and transfer to the delivery location. During transport water needs to be maintained to a high quality, with regulated temperature, sufficient oxygen and minimal waste products. In some cases anaesthetics may be used in small doses to calm fish before transport.

Aquaculture is sometimes part of an environmental rehabilitation program or as an aid in conserving endangered species.

Prospects

Global wild fisheries are in decline, with valuable habitat such as estuaries in critical condition. The aquaculture or farming of piscivorous fish, like salmon, does not help the problem because they need to eat products from other fish, such as fish meal and fish oil. Studies have shown that salmon farming has major negative impacts on wild salmon, as well as the forage fish that need to be caught to feed them. Fish that are higher on the food chain are less efficient sources of food energy.

Apart from fish and shrimp, some aquaculture undertakings, such as seaweed and filter-feeding bivalve mollusks like oysters, clams, mussels and scallops, are relatively benign and even environmentally restorative. Filter-feeders filter pollutants as well as nutrients from the water, improving water quality. Seaweeds extract nutrients such as inorganic nitrogen and phosphorus directly from the water, and filter-feeding mollusks can extract nutrients as they feed on particulates, such as phytoplankton and detritus.

Some profitable aquaculture cooperatives promote sustainable practices. New methods lessen the risk of biological and chemical pollution through minimizing fish stress, fallowing netpens, and applying Integrated Pest Management. Vaccines are being used more and more to reduce antibiotic use for disease control.

Onshore recirculating aquaculture systems, facilities using polyculture techniques, and properly sited facilities (for example, offshore areas with strong currents) are examples of ways to manage negative environmental effects.

Recirculating aquaculture systems (RAS) recycle water by circulating it through filters to remove fish waste and food and then recirculating it back into the tanks. This saves water and the waste gathered can be used in compost or, in some cases, could even be treated and used on land. While RAS was developed with freshwater fish in mind, scientist associated with the Agricultural Research Service have found a way to rear saltwater fish using RAS in low-salinity waters. Although saltwater fish are raised in off-shore cages or caught with nets in water that typically has a salinity of 35 parts per thousand (ppt), scientists were able to produce healthy pompano, a saltwater fish, in tanks with a salinity of only 5 ppt. Commercializing low-salinity RAS are predicted to have positive environmental and economical effects. Unwanted nutrients from the fish food would not be added to the ocean and the risk of transmitting diseases between wild and farm-raised fish would greatly be reduced. The price of expensive saltwater fish, such as the pompano and combia used in the experiments, would be reduced. However, before any of this can be done researchers must study every aspect of the fish's lifecycle, including the amount of ammonia and nitrate the fish will tolerate in the water, what to feed the fish during each stage of its lifecycle, the stocking rate that will produce the healthiest fish, etc.

Some 16 countries now use geothermal energy for aquaculture, including China, Israel, and the United States. In California, for example, 15 fish farms produce tilapia, bass, and catfish with warm water from underground. This warmer water enables fish to grow all year round and mature more quickly. Collectively these California farms produce 4.5 million kilograms of fish each year.

References

- Duram, Leslie A. (2010). Encyclopedia of Organic, Sustainable, and Local Food. ABC-CLIO. p. 139. ISBN 0-313-35963-6.

- G. Mokhtar (1981-01-01). Ancient civilizations of Africa. Unesco. International Scientific Committee for the Drafting of a General History of Africa. p. 309. ISBN 9780435948054. Retrieved 2012-06-19

- Frenken, K. (2005). Irrigation in Africa in figures – AQUASTAT Survey – 2005 (PDF). Food and Agriculture Organization of the United Nations. ISBN 92-5-105414-2. Retrieved 2007-03-14.

- Drainage Manual: A Guide to Integrating Plant, Soil, and Water Relationships for Drainage of Irrigated Lands. Interior Dept., Bureau of Reclamation. 1993. ISBN 0-16-061623-9.

- Bleasdale, J. K. A.; Salter, Peter John (1 January 1991). The Complete Know and Grow Vegetables. Oxford University Press. ISBN 978-0-19-286114-6.

- Ross, Merrill A.; Lembi, Carole A. (2008). Applied Weed Science: Including the Ecology and Management of Ivasive Plants. Prentice Hall. p. 123. ISBN 978-0135028148.

- "Health and Consumer Protection - Scientific Committee on Animal Health and Animal Welfare - Previous outcome of discussions (Scientific Veterinary Committee) - 17". Retrieved September 6, 2015.

- "Mercy For Animals – World's Leading Farmed Animal Rights and Vegan Advocacy Organization - Mercy For Animals". Mercy For Animals. December 17, 2014. Retrieved September 6, 2015.

- Written Jennifer Ackerman. "Food Article, Foodborne Illness Information, Pathogen Facts -- National Geographic". National Geographic. Retrieved September 6, 2015..

- Food Standards Agency. "[ARCHIVED CONTENT] Food Standards Agency - VPC report on growth hormones in meat". Retrieved September 6, 2015..

- Undersander, Dan. "Pastures for Profit, a guide to rotational grazing" (PDF). USDA-NRCS. University of Minnesota extension service. Retrieved 10 December 2015.

Agricultural Machinery and its Types

Agricultural machinery is the machinery that is used in either farming or any other agricultural practice. Some of the types of machines explained in this section are manure spreaders, cultivators, ploughs, drip irrigation, hog oiler and bulk tanks. The topics discussed in the section are of great importance to broaden the existing knowledge on agricultural machinery.

Agricultural Machinery

Agricultural machinery is machinery used in farming or other agriculture. There are many types of such equipment, from hand tools and power tools to tractors and the countless kinds of farm implements that they tow or operate. Diverse arrays of equipment are used in both organic and nonorganic farming. Especially since the advent of mechanised agriculture, agricultural machinery is an indispensable part of how the world is fed.

A German combine harvester by Claas

History of the Machines

The Industrial Revolution

With the coming of the Industrial Revolution and the development of more complicated machines, farming methods took a great leap forward. Instead of harvesting grain by hand with a sharp blade, wheeled machines cut a continuous swath. Instead of threshing the grain by beating it with sticks, threshing machines separated the seeds from the heads and stalks. The first tractors appeared in the late 19th century.

Steam Power

Power for agricultural machinery was originally supplied by ox or other domesticated animals. With the invention of steam power came the portable engine, and later the traction engine, a multipurpose, mobile energy source that was the ground-crawling cousin to the steam locomotive. Agricultural steam engines took over the heavy pulling work of oxen, and were also equipped with a pulley that could power stationary machines via the use of a long belt. The steam-powered machines were low-powered by today's standards but, because of their size and their low gear ratios, they could provide a large drawbar pull. Their slow speed led farmers to comment that tractors had two speeds: "slow, and damn slow."

Internal Combustion Engines

The internal combustion engine; first the petrol engine, and later diesel engines; became the main source of power for the next generation of tractors. These engines also contributed to the development of the self-propelled, combined harvester and thresher, or combine harvester (also shortened to 'combine'). Instead of cutting the grain stalks and transporting them to a stationary threshing machine, these combines cut, threshed, and separated the grain while moving continuously through the field.

Types

A John Deere cotton harvester at work in a cotton field.

Combines might have taken the harvesting job away from tractors, but tractors still do the majority of work on a modern farm. They are used to push implements—machines that till the ground, plant seed, and perform other tasks.

Tillage implements prepare the soil for planting by loosening the soil and killing weeds or competing plants. The best-known is the plow, the ancient implement that was upgraded in 1838 by John Deere. Plows are now used less frequently in the U.S. than formerly, with offset disks used instead to turn over the soil, and chisels used to gain the depth needed to retain moisture.

The most common type of seeder is called a planter, and spaces seeds out equally in long rows, which are usually two to three feet apart. Some crops are planted by drills, which put out much more seed in rows less than a foot apart, blanketing the field with crops. Transplanters automate the task of transplanting seedlings to the field. With the widespread use of plastic mulch, plastic mulch layers, transplanters, and seeders lay down long rows of plastic, and plant through them automatically.

From left to right: John Deere 7800 tractor with Houle slurry trailer, Case IH combine harvester, New Holland FX 25 forage harvester with corn head.

After planting, other implements can be used to cultivate weeds from between rows, or to spread fertilizer and pesticides. Hay balers can be used to tightly package grass or alfalfa into a storable form for the winter months.

Modern irrigation relies on machinery. Engines, pumps and other specialized gear provide water quickly and in high volumes to large areas of land. Similar types of equipment can be used to deliver fertilizers and pesticides.

Besides the tractor, other vehicles have been adapted for use in farming, including trucks, airplanes, and helicopters, such as for transporting crops and making equipment mobile, to aerial spraying and livestock herd management.

New Technology and the Future

The basic technology of agricultural machines has changed little in the last century. Though modern harvesters and planters may do a better job or be slightly tweaked from their predecessors, the US$250,000 combine of today still cuts, threshes, and separates grain in the same way it has always been done. However, technology is changing the way that humans operate the machines, as computer monitoring systems, GPS locators, and self-steer programs allow the most advanced tractors and implements to be more precise and less wasteful in the use of fuel, seed, or fertilizer. In the foreseeable future, there may be mass production of driverless tractors, which use GPS maps and electronic sensors.

Open Source Agricultural Equipment

Many farmers are upset by their inability to fix the new types of high-tech farm equipment. This is due mostly to companies using intellectual property law to prevent farmers from having the legal right to fix their equipment (or gain access to the information to allow them to do it). This has encouraged groups such as Open Source Ecology and Farm Hack to begin to make open source agricultural machinery. In addition on a smaller scale FarmBot and the RepRap open source 3D printer community has begun to make open-source farm tools available of increasing levels of sophistication. In October 2015 an exemption was added to the DMCA to allow inspection and modification of the software in cars and other vehicles including agricultural machinery.

Mechanised Agriculture

A cotton picker at work. The first successful models were introduced in the mid-1940s and each could do the work of 50 hand pickers.

Mechanised agriculture is the process of using agricultural machinery to mechanise the work of agriculture, greatly increasing farm worker productivity. In modern times, powered machinery has replaced many farm jobs formerly carried out by manual labour or by working animals such as oxen, horses and mules.

The entire history of agriculture contains many examples of the use of tools, such as the hoe and the plough. But the ongoing integration of machines since the Industrial Revolution has allowed farming to become much less labour-intensive.

Current mechanised agriculture includes the use of tractors, trucks, combine harvesters, countless types of farm implements, aeroplanes and helicopters (for aerial application), and other vehicles. Precision agriculture even uses computers in conjunction with satellite imagery and satellite navigation (GPS guidance) to increase yields.

Mechanisation was one of the large factors responsible for urbanisation and industrial economies. Besides improving production efficiency, mechanisation encourages large scale production and sometimes can improve the quality of farm produce. On the other hand, it can displace unskilled farm labour and can cause environmental degradation (such as pollution, deforestation, and soil erosion), especially if it is applied shortsightedly rather than holistically.

History

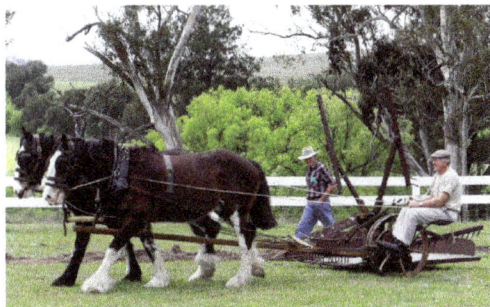

A reaper at Woolbrook, New South Wales

Threshing machine in 1881. Steam engines were also used to power threshing machines. Today both reaping and threshing are done with a combine harvester.

Jethro Tull's seed drill (ca. 1701) was a mechanical seed spacing and depth placing device that increased crop yields and saved seed. It was an important factor in the British Agricultural Revolution.

Since the beginning of agriculture threshing was done by hand with a flail, requiring a great deal of labour. The threshing machine, which was invented in 1794 but not widely used for several more decades, simplified the operation and allowed the use of animal power. Before the invention of the grain cradle (ca. 1790) an able bodied labourer could reap about one quarter acre of wheat in a day using a sickle. It was estimated that for each of Cyrus McCormick's horse pulled reapers (ca. 1830s) freed up five men for military service in the US Civil War. Later innovations included raking and binding machines. By 1890 two men and two horses could cut, rake and bind 20 acres of wheat per day.

In the 1880s the reaper and threshing machine were combined into the combine harvester. These machines required large teams of horses or mules to pull. Steam power was applied to threshing machines in the late 19th century. There were steam engines that moved around on wheels under their own power for supplying temporary power to stationary threshing machines. These were called *road engines,* and Henry Ford seeing one as a boy was inspired to build an automobile.

With internal combustion came the first modern tractors in the early 1900s, becoming more popular after the Fordson tractor (ca. 1917). At first reapers and combine harvesters were pulled by tractors, but in the 1930s self powered combines were developed.

Advertising for motorised equipment in farm journals during this era did its best to compete against horse-drawn methods with economic arguments, extolling common themes such as that a tractor "eats only when it works", that one tractor could replace many horses, and that mechanisation could allow one man to get more work done per day than he ever had before. The horse population in the US began to decline in the 1920s after the conversion of agriculture and transportation to internal combustion. Peak tractor sales in the US were around 1950. In addition to saving labour, this freed up much land previously used for supporting draft animals. The greatest period of growth in agricultural productivity in the US was from the 1940s to the 1970s, during which time agriculture was benefiting from internal combustion powered tractors and combine harvesters, chemical fertilizers and the green revolution.

Although farmers of corn, wheat, soy, and other commodity crops had replaced most of their workers with harvesting machines and combines enabling them to efficiently cut and gather grains, growers of produce continued to rely on human pickers to avoid the bruising of the product in order to maintain the blemish-free appearance demanded of consumers. The continuous supply of illegal workers from Latin America that were willing to harvest the crops for low wages further suppressed the need for mechanisation. As the number of illegal workers has continued to decline since reaching its peak in 2007 due to increased border patrols and an improving Mexican economy, the industry is increasing the use of mechanisation. Proponents argue that mechanisation will boost productivity and help to maintain low food prices while farm worker advocates assert that it will eliminate jobs and will give an advantage to large growers who are able to afford the required equipment.

Current Status of Future Applications

Asparagus Harvesting

Asparagus are presently harvested by hand with labour costs at 71% of production costs and 44% of selling costs. Asparagus is a difficult crop to harvest since each spear matures at a different speed making it difficult to achieve a uniform harvest. A prototype asparagus harvesting machine - using a light-beam sensor to identify the taller spears - is expected to be available for commercial use.

Blueberry Harvesting

Mechanization of Maine's blueberry industry has reduced the number of migrant workers required from 5,000 in 2005 to 1,500 in 2015 even though production has increased from 50-60 million pounds per year in 2005 to 90 million pounds in 2015.

Chili Pepper Harvesting

As of 2014, prototype chili pepper harvesters are being tested by New Mexico State University. The New Mexico green chile crop is currently hand-picked entirely by field workers as chili pods tend to bruise easily. The first commercial application commenced in 2015. The equipment is expected to increase yield per acre and help to offset a sharp decline in acreage planted due to the lack of available labour and drought conditions.

Orange Harvesting

As of 2010, approximately 10% of the processing orange acreage in Florida is harvested mechanically. Mechanisation has progressed slowly due to the uncertainty of future economic benefits due to competition from Brazil and the transitory damage to orange trees when they are harvested.

Raisin Harvesting

As of 2007, mechanised harvesting of raisins is at 45%; however the rate has slowed due to high raisin demand and prices making the conversion away from hand labour less urgent. A new strain of grape developed by the USDA that drys on the vine and is easily harvested mechanically is expected to reduce the demand for labour.

Strawberry Harvesting

Strawberries are a high cost-high value crop with the economics supporting mechanisation. In 2005, picking and hauling costs were estimated at $594 per ton or 51% of the total grower cost. However, the delicate nature of fruit make it an unlikely candidate for mechanisation in the near future. A strawberry harvester developed by Shibuya Seiki and unveiled in Japan in 2013 is able to pick a strawberry every eight seconds. The robot identifies which strawberries are ready to pick by using three separate cameras and then once identified as ready, a mechanised arm snips the fruit free and gently places it in a basket. The robot moves on rails between the rows of strawberries which are generally contained within elevated greenhouses. The machine costs 5 million yen. A new strawberry harvester made by Agrobot that will harvest strawberries on raised, hydroponic beds using 60 robotic arms is expected to be released in 2016.

Tomato Harvesting

Mechanical harvesting of tomatoes started in 1965 and as of 2010, nearly all processing tomatoes are mechanically harvested. As of 2010, 95% of the US processed tomato crop is produced in California. Although fresh market tomatoes have substantial hand harvesting costs (in 2007, the costs of hand picking and hauling were $86 per ton which is 19% of total grower cost), packing and selling costs were more of a concern (at 44% of total grower cost) making it likely that cost saving efforts would be applied there.

According to a 1977 report by the California Agrarian Action Project, during the summer of 1976 in California, many harvest machines had been equipped with a photo-electric scanner that sorted out green tomatoes among the ripe red ones using infrared lights and colour sensors. It worked in lieu of 5,000 hand harvesters causing displacement of innumerable farm labourers as well as wage cuts and shorter work periods. Migrant workers were hit the hardest. To withstand the rigour of the machines, new crop varieties were bred to match the automated pickers. UC Davis Professor G.C. Hanna propagated a thick-skinned tomato called VF-145. But even still, millions were damaged with impact cracks and university breeders produced a more tougher and juiceless "square round" tomato. Small farms were of insufficient size to obtain financing to purchase the equipment and within 10 years, 85% of the state's 4,000 cannery tomato farmers were out of the business. This led to a concentrated tomato industry in California that "now packed 85% of the nation's tomato products". The monoculture fields fostered rapid pest growth, requiring the use of "more than four million pounds of pesticides each year" which greatly affected the health of the soil, the farm workers, and possibly the consumers.

Manure Spreader

A modern manure spreader

A manure spreader or muck spreader or honey wagon is an agricultural machine used to distribute manure over a field as a fertilizer. A typical (modern) manure spreader consists of a trailer towed behind a tractor with a rotating mechanism driven by the tractor's power take off (PTO). Truck mounted manure spreaders are also common in North America.

Operation

Manure spreaders began as ground-driven units which could be pulled by a horse or team of horses. Many of these ground-driven spreaders are still produced today, mostly in the form of small units that can be pulled behind a larger garden tractor or an all terrain vehicle (ATV). In recent years hydraulic and PTO driven units have been developed to offer variable application rates. Several models are also designed with removable rotating mechanisms (beaters), attachable side extensions, and tailgates for hauling chopped forages, cereal grains, and other crops. A typical (modern) manure spreader consists of a trailer towed behind a tractor with a rotating mechanism driven by the tractor's power take off (PTO).

History

An advertisement for a J.S.Kemp model spreader

The first successful automated manure spreader was designed by Joseph Kemp in 1875. Manure spreaders began as ground-driven units which could be pulled by a horse or team of horses. At the time of his invention, he was living in Waterloo, Canada, but thereafter, he moved to Newark Valley, NY and formed the J.S. Kemp Manufacturing Co. to manufacture and market his current and subsequent designs. In 1903, he expanded the company to Waterloo, Iowa before selling the design to International Harvester, in 1906.

The original New Idea spreader design. Note the paddle system at the rear that creates the 'widespreading' effect.

Joseph Oppenheim of Maria Stein, Ohio was the inventor of the first modern 'widespreading' manure spreader and is honored as such in the Ohio Agricultural Hall of Fame. Originally manure was thrown from a wagon. Later, "manure unloaders" used a drag chain at the bottom of the wagon to pull the load of manure to the rear where it was shredded by a pair of beaters. Because the unloaders deposited manure directly behind the wagon but with very little spreading to the sides, farmers still had to take the time-consuming step of heading into the fields with peg-tooth drags or similar implements to spread the manure in order to prevent burning the soil.

Oppenheim, a schoolmaster in the small town, concerned that his older male students often missed school loading and spreading manure, patented a wagon that, behind the drag chain and two beaters, incorporated a steel axle with several wooden paddles attached to the shaft at an angle to throw the manure outward in a broad pattern eliminating the necessity for manual spreading. On October 18, 1899, Oppenheim began to produce his new manure spreader, incorporating the "widespread" paddle device. Neighbors soon referred to it as "Oppenheim's new idea" and Oppenheim adopted this name for his business.

Although Oppenheim died in November, 1901, the demand for the New Idea Spreader Company's labor-saving "widespread" machines quickly grew and fifteen years later, under the direction of his oldest son, B.C. Oppenheim, and Henry Synck, one of Oppenheim's first employees, the company, had branches in eight states and an assembly plant in Guelph, Ontario. It had total sales in 1916 of $1,250,000. Eight years later, in 1924, the factory was turning out 125 manure spreaders in an eight-hour day. and "became the brand that set the standards for spreader performance, durability and reliability decade after decade."

During the 1920s, Henry Synck, who became president of the company in 1936 when Joseph's son, B.C. Oppenheim, died, patented several improvements to the spreader. In 1945 the Oppenheim family sold its controlling interest in the closely held New Idea Company to AVCO Manufacturing. AVCO later sold the company to White Farm Equipment Company which in 1993 sold it to AGCO (Allis-Gleaner Corporation), the current owner.

It is clear, however, that there were other competitors in this field, each of whom spread the manure by a slightly different technique. One of these is the Great Western Farm Equipment Line, produced in Chicago, IL.

In Popular Culture

In the post-Civil War AMC TV series *Hell on Wheels*, the muck wagon is a vehicle used to collect dead bodies, animal carcasses, and filled chamber pots. When The Swede is disgraced, Durant demotes him from head of security to driver of the muck wagon.

Cultivator

A cultivator is any of several types of farm implement used for secondary tillage. One sense of the name refers to frames with the teeth (also called shanks) that pierce the soil as they are dragged through it linearly. Another sense refers to machines that use rotary motion of disks or teeth to accomplish a similar result. The rotary tiller is a principle example.

A cultivator pulled by a tractor in Canada in 1943

Cultivators stir and pulverize the soil, either before planting (to aerate the soil and prepare a smooth, loose seedbed) or after the crop has begun growing (to kill weeds—controlled disturbance of the topsoil close to the crop plants kills the surrounding weeds by uprooting them, burying their leaves to disrupt their photosynthesis, or a combination of both). Unlike a harrow, which disturbs the entire surface of the soil, cultivators are designed to disturb the soil in careful patterns, sparing the crop plants but disrupting the weeds.

Cultivators of the toothed type are often similar in form to chisel plows, but their goals are different. Cultivator teeth work near the surface, usually for weed control, whereas chisel plow shanks work deep beneath the surface, breaking up hardpan. Consequently, cultivating also takes much less power per shank than does chisel plowing.

Small toothed cultivators pushed or pulled by a single person are used as garden tools for small-scale gardening, such as for the household's own use or for small market gardens. Similarly sized rotary tillers combine the functions of harrow and cultivator into one multipurpose machine.

Cultivators are usually either self-propelled or drawn as an attachment behind either a two-wheel tractor or four-wheel tractor. For two-wheel tractors they are usually rigidly fixed and powered via couplings to the tractors' transmission. For four-wheel tractors they are usually attached by means of a three-point hitch and driven by a power take-off (PTO). Drawbar hookup is also still commonly used worldwide. Draft-animal power is sometimes still used today, being somewhat common in developing nations although rare in more industrialized economies.

History

The basic idea of soil scratching for weed control is ancient and was done with hoes or mattocks for millennia before cultivators were developed. Cultivators were originally drawn by draft animals (such as horses, mules, or oxen) or were pushed or drawn by people. In modern commercial agriculture, the amount of cultivating done for weed control has been greatly reduced via use of herbicides instead. However, herbicides are not always desirable—for example, in organic farming.

The powered rotary hoe was invented by Arthur Clifford Howard who, in 1912, began experimenting with rotary tillage on his father's farm at Gilgandra, New South Wales, Australia. Initially using his father's steam tractor engine as a power source, he found that ground could be mechanically tilled without soil-packing occurring, as was the case with normal ploughing. His earliest designs threw the tilled soil sideways, until he improved his invention by designing an L-shaped blade mounted on widely spaced flanges fixed to a small-diameter rotor. With fellow apprentice Everard McCleary, he established a company to make his machine, but plans were interrupted by World War I. In 1919 Howard returned to Australia and resumed his design work, patenting a design with 5 rotary hoe cultivator blades and an internal combustion engine in 1920.

In March 1922, Howard formed the company Austral Auto Cultivators Pty Ltd, which later became known as Howard Auto Cultivators. It was based in Northmead, a suburb of Sydney, from 1927.

Meanwhile, in North America during the 1910s, tractors were evolving away from traction engine-sized monsters toward smaller, lighter, more affordable machines. The Fordson tractor especially had made tractors affordable and practical for small and medium family farms for the first time in history. Cultivating was somewhat of an afterthought in the Fordson's design, which reflected the fact that even just bringing practical motorized tractive power alone to this market segment was in itself a milestone. This left an opportunity for others to pursue better motorized cultivating. Between 1915 and 1920, various inventors and farm implement companies experimented with a class of machines referred to as *motor cultivators*, which were simply modified horse-drawn shank-type cultivators with motors added for self-propulsion. This class of machines found limited market success. But by 1921 International Harvester had combined motorized cultivating with the other tasks of tractors (tractive power and belt work) to create the Farmall, the general-purpose tractor tailored to cultivating that basically invented the category of row-crop tractors.

In Australia, by the 1930s, Howard was finding it increasingly difficult to meet a growing worldwide demand for exports of his machines. He travelled to the United Kingdom, founding the company Rotary Hoes Ltd in East Horndon, Essex, in July 1938. Branches of this new company subsequently opened in the United States of America, South Africa, Germany, France, Italy, Spain, Brazil, Malaysia, Australia and New Zealand. It later became the holding company for Howard Rotavator Co. Ltd. The Howard Group of companies was acquired by the Danish Thrige Agro Group in 1985, and in December 2000 the Howard Group became a member of Kongskilde Industries of Soroe, Denmark.

When herbicidal weed control was first widely commercialized in the 1950s and 1960s, it played into that era's optimistic worldview in which sciences such as chemistry would usher in a new age of modernity that would leave old-fashioned practices (such as weed control via cultivators) in the dustbin of history. Thus herbicidal weed control was adopted very widely, and in some cases too heavily and hastily. In subsequent decades, people overcame this initial imbalance and came to realize that herbicidal weed control has limitations and externalities, and it must be managed intelligently. It is still widely used, and probably will continue to be indispensable to affordable food production worldwide for the foreseeable future; but its wise management includes seeking alternate methods, such as the traditional standby of mechanical cultivation, where practical.

Industrial Use

To the extent that cultivating is done commercially today (such as in truck farming), it is usually powered by tractors, especially row-crop tractors. Industrial cultivators can vary greatly in size and shape, from 10 feet (3 m) to 80 feet (24 m) wide. Many are equipped with hydraulic wings that fold up to make road travel easier and safer. Different types are used for preparation of fields before planting, and for the control of weeds between row crops. The cultivator may be an implement trailed after the tractor via a drawbar; mounted on the three-point hitch; or mounted on a frame beneath the tractor. Active cultivator implements are driven by a power take-off shaft. While most cultivator are considered a secondary tillage implement, active cultivators are commonly used for primary tillage in lighter soils instead of plowing. The largest versions available are about 6 m (20 ft) wide, and require a tractor with an excess of 150 horsepower (110 kW) (PTO) to drive them.

Field cultivators are used to complete tillage operations in many types of arable crop fields. The main function of the field cultivator is to prepare a proper seedbed for the crop to be planted into, to bury crop residue in the soil (helping to warm the soil before planting), to control weeds, and to mix and incorporate the soil to ensure the growing crop has enough water and nutrients to grow well during the growing season. The implement has many shanks mounted on the underside of a metal frame, and small narrow rods at the rear of the machine that smooth out the soil surface for easier travel later when planting. In most field cultivators, one-to-many hydraulic cylinders raise and lower the implement and control its depth.

Row Crop Cultivators

Home made sweep. Notice the inner and outer "sweep" blades.

The main function of the row crop cultivator is weed control between the rows of an established crop. Row crop cultivators are usually raised and lowered by a three-point hitch and the depth is controlled by gauge wheels.

Sometimes referred to as *sweep cultivators*, these commonly have two center blades that cut weeds from the roots near the base of the crop and turn over soil, while two rear sweeps further outward than the center blades deal with the center of the row, and can be anywhere from 1 to 36 rows wide.

Garden Cultivators

Small tilling equipment, used in small gardens such as household gardens and small commercial gardens, can provide both primary and secondary tillage. For example, a rotary tiller does both the "plowing" and the "harrowing", preparing a smooth, loose seedbed. It does not provide the row-wise weed control that cultivator teeth would. For that task, there are single-person-pushable toothed cultivators.

Variants and Trademarks

A Japanese two-wheel tractor

Rotary tillers are a type of cultivators. Rotary tillers are popular with home gardeners who want large vegetable gardens. The garden may be tilled a few times before planting each crop. Rotary tillers may be rented from tool rental centers for single-use applications, such as when planting grass.

A small rotary hoe for domestic gardens was known by the trademark Rototiller and another, made by the Howard Group, who produced a range of rotary tillers, was known as the Rotavator.

Rototiller

> The small rototiller is typically propelled forward via a (1–5 horsepower or 0.8–3.5 kilowatts) petrol engine rotating the tines, and do not have powered wheels, though they may have small transport/level control wheel(s). To keep the machine from moving forward too fast, an adjustable tine is usually fixed just behind the blades so that through friction with deeper un-tilled soil, it acts

as a brake, slowing the machine and allowing it to pulverize the soils. The slower a rototiller moves forward, the more soil tilth can be obtained. The operator can control the amount of friction/braking action by raising and lowering the handlebars of the tiller. Rototillers do not have a reverse as such backwards movement towards the operator could cause serious injury. While operating, the rototiller can be pulled backwards to go over areas that were not pulverized enough, but care must be taken to ensure that the operator does not stumble and pull the rototiller on top of himself. Rototilling is much faster than manual tilling, but notoriously difficult to handle and exhausting work, especially in the heavier and higher horsepower models. If the rototiller's blades catch on unseen subsurface objects, such as tree roots and buried garbage, it can cause the rototiller to abruptly and violently move in any direction.

Rotavator

Unlike the Rototiller, the self-propelled Howard Rotavator is equipped with a gearbox and driven forward, or held back, by its wheels. The gearbox enables the forward speed to be adjusted while the rotational speed of the tines remains constant which enables the operator to easily regulate the extent to which soil is engaged. For a two-wheel tractor rotavator this greatly reduces the workload of the operator as compared to a rototiller. These rotavators are generally more heavy duty, come in higher power (4–18 horsepower or 3–13 kilowatts) with either petrol or diesel engines and can cover much more area per hour. The trademarked word "Rotavator" is one of the longest single-word palindromes in the English language.

Mini tiller

Mini tillers are a new type of small agricultural tillers or cultivators used by farmers or homeowners. These are also known as power tillers or garden tillers. Compact, powerful and, most importantly, inexpensive, these agricultural rotary tillers are providing alternatives to four-wheel tractors and in the small farmers' fields in developing countries are more economical than four-wheel tractors.

Two-wheel tractor

The higher power "riding" rotavators cross out of the home garden category into farming category, especially in Asia, Africa and South America, capable of preparing 1 hectare of land in 8–10 hours. These are also known as *power tillers* or *walking tractors*. Years ago they were considered only useful for rice growing areas, where they were fitted with steel cage-wheels for traction, but now the same are being used in both wetland and dryland farming all over the world. They have multiple functions with related tools for dryland or paddys, pumping, transportation, threshing, ditching, spraying pesticide. They can be used on hills, mountains, in greenhouses and orchards. Diesel designs are more popular in developing countries than gasoline.

Harrow (Tool)

A spring-tooth drag harrow

In agriculture, a harrow (often called a set of harrows in a plurale tantum sense) is an implement for breaking up and smoothing out the surface of the soil. In this way it is distinct in its effect from the plough, which is used for deeper tillage. Harrowing is often carried out on fields to follow the rough finish left by plowing operations. The purpose of this harrowing is generally to break up clods (lumps of soil) and to provide a finer finish, a good tilth or soil structure that is suitable for seedbed use. Coarser harrowing may also be used to remove weeds and to cover seed after sowing. Harrows differ from cultivators in that they disturb the whole surface of the soil, such as to prepare a seedbed, instead of disturbing only narrow trails that skirt crop rows (to kill weeds).

Disk Harrow, p. 641.

Disc harrows

There are four general types of harrows: disc harrows, tine harrows (including spring-tooth harrows, drag harrows, and spike harrows), chain harrows, and chain-disk harrows. Harrows were originally drawn by draft animals, such as horses, mules, or oxen, or in some times and places by manual labourers. In modern practice they are almost always tractor-mounted implements, either trailed after the tractor by a drawbar or mounted on the three-point hitch.

Crumbler roller, commonly used to compact soil after it has been loosened by a harrow

Clydesdale horses pulling spike harrows, Murrurundi, NSW, Australia

A modern development of the traditional harrow is the rotary power harrow, often just called a power harrow.

Types

In cooler climates the most common types are the *disc harrow*, the *chain harrow*, the *tine harrow* or *spike harrow* and the *spring tine harrow*. Chain harrows are often used for lighter work such as levelling the tilth or covering seed, while disc harrows are typically used for heavy work, such as following ploughing to break up the sod. In addition, there are various types of *power harrow*, in which the cultivators are power-driven from the tractor rather than depending on its forward motion.

Tine harrows are used to refine seed-bed condition before planting, to remove small weeds in growing crops and to loosen the inter-row soils to allow for water to soak into the subsoil. The fourth is a chain disk harrow. Disk attached to chains are pulled at an angle over the ground. These harrows move rapidly across the surface. The chain and disk rotate to stay clean while breaking up the top surface to about 1 inch (3 cm) deep. A smooth seedbed is prepared for planting with one pass.

Harrowing with tractor and disk harrow in the 1940s)

Chain harrowing can be used on pasture land to spread out dung, and to break up dead material (*thatch*) in the sward, and similarly in sports-ground maintenance a light chain harrowing is often used to level off the ground after heavy use, to remove and smooth out boot marks and indentations. Used on tilled land in combination with the other two types, chain harrowing rolls remaining larger soil clumps to the surface where weather breaks them down and prevents interference with seed germination.

All four harrow types can be used in one pass to prepare soil for seeding. It is also common to use any combination of two harrows for a variety of tilling processes. Where harrowing provides a very fine tilth, or the soil is very light so that it might easily be wind-blown, a roller is often added as the last of the set.

Harrows may be of several types and weights, depending on their purpose. They almost always consist of a rigid frame that holds discs, teeth, linked chains, or other means of moving soil—but tine and chain harrows are often only supported by a rigid towing-bar at the front of the set.

In the southern hemisphere, so-called *giant discs* are a specialised kind of disc harrows that can stand in for a plough in rough country where a mouldboard plough cannot handle tree-stumps and rocks, and a disc-plough is too slow (because of its limited number of discs). Giant scalloped-edged discs operate in a set, or frame, that is often weighted with concrete or steel blocks to improve penetration of the cutting edges. This sort of cultivation is usually followed by broadcast fertilisation and seeding, rather than drilled or row seeding.

A drag is a heavy harrow.

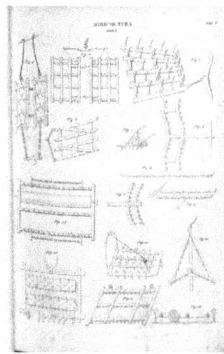

19th century harrow

Power Harrow

A rotary power harrow, or simply power harrow, has multiple sets of vertical tines. Each set of tines is rotated on a vertical axis and tills the soil horizontally. The result is that, unlike a rotary tiller, soil layers are not turned over or inverted, which is useful in preventing dormant weed seeds from being brought to the surface, and there is no horizontal slicing of the subsurface soil that can lead to hardpan formation.

Historical Reference

An Arabic reference to harrows is to be found in Abu Bakr Ibn Wahshiyya's Nabatean

Agriculture (Kitab al-Filaha al-Nabatiyya), of the 10th century, but claiming knowledge from Babylonian sources. In Europe, harrows were first used in the Middle Ages.

The oldest known illustration of a harrow is in Scene 10 of the eleventh-century Bayeux Tapestry.

Spike harrow depicted on a 16th-century German coat-of-arms

Plough

Traditional ploughing: a farmer works the land with horses and plough

A plough (UK) or plow tool or farm implement used in farming for initial cultivation of soil in preparation for sowing seed or planting to loosen or turn the soil. Ploughs are traditionally drawn by working animals such as horses or cattle, but in modern times may be drawn by tractors. A plough may be made of wood, iron, or steel frame with an attached blade or stick used to cut the earth. It has been a basic instrument for most of recorded history, although written references to the plough do not appear in English until c. 1100 at which point it is referenced frequently. The plough represents one of the major agricultural inventions in human history.

The primary purpose of ploughing is to turn over the upper layer of the soil, bringing fresh nutrients to the surface, while burying weeds and the remains of previous crops and allowing them to break down. As the plough is drawn through the soil it creates long trenches of fertile soil called furrows. In modern use, a ploughed field is typically left to dry out, and is then harrowed before planting. Plowing and cultivating a soil homogenizes and modifies the upper 12 to 25 cm of the soil to form a plow layer. In many soils, the majority of fine plant feeder roots can be found in the topsoil or plow layer.

Modern tractor ploughing in South Africa. This plough has five non-reversible mouldboards. The fifth, empty furrow on the left may be filled by the first furrow of the next pass.

Ploughing with oxen. A miniature from an early-sixteenth-century manuscript of the Middle English poem *God Spede þe Plough*, held at the British Museum

Oxen driven ploughing

Ploughs were initially human powered, but the process became considerably more efficient once animals were pressed into service. The first animal powered ploughs were undoubtedly pulled by oxen, and later in many areas by horses (generally draft horses) and mules, although various other animals have been used for this purpose. In industrialised countries, the first mechanical means of pulling a plough were steam-powered (ploughing engines or steam tractors), but these were gradually superseded by internal-combustion-powered tractors.

Modern competitions take place for ploughing enthusiasts like the National Ploughing Championships in Ireland. Use of the plough has decreased in some areas, often those significantly threatened by soil damage and erosion, in favour of shallower ploughing and other less invasive conservation tillage techniques.

Natural farming methods are emerging that do not involve any ploughing at all, unless an initial ploughing is necessary to break up hardpan on a new plot to be cultivated, so that the newly introduced soil life can penetrate and develop more quickly and deeply. By not ploughing, beneficial fungi and microbial life can develop that will eventually bring air into the soil, retain water and build up nutrients. A healthy soil full of active fungi and microbial life, combined with a diverse crop (making use of companion planting), suppresses weeds and pests naturally and retains rainwater. Thus the intensive use of water-, oil- and energy hungry irrigation, fertilizers and herbicides are avoided. Cultivated land becomes more fertile and productive over time, while tilled land tends to go down in productivity over time due to erosion and the removal of nutrients with every harvest. Proponents of permaculture claim that it is the only way of farming that can be maintained when fossil fuel runs out. On the other hand, the advantage of agricultural methods that require repeated ploughing are that they allow monocropping on a large scale at remote locations, using industrial machinery rather than human labor.

Etymology

Ploughing.

In older English, as in other Germanic languages, the plough was traditionally known by other names, e.g. Old English *sulh*, Old High German *medela*, *geiza*, *huohili(n)*, Old Norse *arðr* (Swedish *årder*), and Gothic *hōha*, all presumably referring to the ard (scratch plough). The term plough or plow, as used today, was not common until 1700.

The modern word *plough* comes from Old Norse *plógr*, and therefore Germanic, but it appears relatively late (it is not attested in Gothic), and is thought to be a loanword from one of the north Italic languages. Words with the same root appeared with related meanings: in Raetic *plaumorati* "wheeled heavy plough" (Pliny, *Nat. Hist.* 18, 172), and in Latin *plaustrum* "farm cart", *plōstrum*, *plōstellum* "cart", and *plōxenum*, *plōximum* "cart box". The word must have originally referred to the wheeled heavy plough, which was common in Roman northwestern Europe by the A.D. 5th century.

Orel (2003) tentatively attaches *plough* to a PIE stem *blōkó-, which gave Armenian *pe☐em* "to dig" and Welsh *bwlch* "crack", though the word may not be of Indo-European origin.

Parts

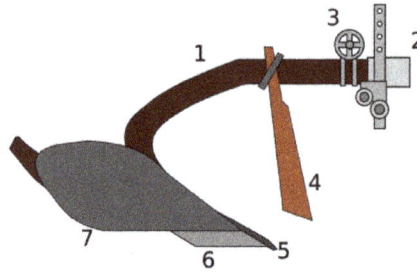

Diagram - modern plough

The diagram (*right*) shows the basic parts of the modern plough:

1. beam

2. hitch (Brit: hake)

3. vertical regulator

4. coulter (knife coulter pictured, but disk coulter common)

5. chisel (foreshare)

6. share (mainshare)

7. moldboard

Other parts not shown or labelled include the frog (or frame), runner, landside, shin, trashboard, and stilts (handles).

On modern ploughs and some older ploughs, the mouldboard is separate from the share and runner, so these parts can be replaced without replacing the mouldboard. Abrasion eventually destroys all parts of a plough that come into contact with the soil.

History

Ploughing with buffalo in Hubei, China

Hoeing

When agriculture was first developed, simple hand-held digging sticks and hoes were used in highly fertile areas, such as the banks of the Nile where the annual flood rejuvenates the soil, to create drills (furrows) to plant seeds in. Digging sticks, hoes, and mattocks were not invented in any one place, and hoe-cultivation must have been common everywhere agriculture was practiced. Hoe-farming is the traditional tillage method in tropical or sub-tropical regions, which are characterized by stony soils, steep slope gradients, predominant root crops, and coarse grains grown at wide distances apart. While hoe-agriculture is best suited to these regions, it is used in some fashion everywhere. Instead of hoeing, some cultures use pigs to trample the soil and grub the earth.

Ard

Ancient Egyptian ard, *c.*1200 BC. (Burial chamber of Sennedjem)

Some ancient hoes, like the Egyptian *mr*, were pointed and strong enough to clear rocky soil and make seed drills, which is why they are called *hand-ards*. However, the domestication of oxen in Mesopotamia and the Indus valley civilization, perhaps as early as the 6th millennium BC, provided mankind with the draft power necessary to develop the larger, animal-drawn true ard (or scratch plough). The earliest was the *bow ard*, which consists of a *draft-pole* (or *beam*) pierced by a thinner vertical pointed stick called the *head* (or *body*), with one end being the *stilt* (handle) and the other a *share* (cutting blade) that was dragged through the topsoil to cut a shallow furrow ideal for most cereal crops. The ard does not clear new land well, so hoes or mattocks must be used to pull up grass and undergrowth, and a hand-held, coulter-like *ristle* could be used to cut deeper furrows ahead of the share. Because the ard leaves a strip of undisturbed earth between the furrows, the fields are often cross-ploughed lengthwise and widthwise, and this tends to form squarish fields (Celtic fields). The ard is best suited to loamy or sandy soils that are naturally fertilized by annual flooding, as in the Nile Delta and Fertile Crescent, and to a lesser extent any other cereal-growing region with light or thin soil. By the late Iron Age, ards in Europe were commonly fitted with coulters.

Mouldboard Plough

Water buffalo used for ploughing in Si Phan Don, Laos

To grow crops regularly in less fertile areas, the soil must be turned to bring nutrients to the surface. A major advance for this type of farming was the turnplough, also known as the mouldboard plough (UK), moldboard plow (US), or frame-plough. A *coulter* (or *skeith*) could be added to cut vertically into the ground just ahead of the *share* (in front of the *frog*), a wedge-shaped cutting edge at the bottom front of the *mouldboard* with the landside of the frame supporting the undershare (below-ground component).

The upper parts of the frame carry (from the front) the coupling for the motive power (horses), the coulter and the landside frame. Depending on the size of the implement, and the number of furrows it is designed to plough at one time, a forecarriage with a wheel or wheels (known as a furrow wheel and support wheel) may be added to support the frame (wheeled plough). In the case of a single-furrow plough there is only one wheel at the front and handles at the rear for the ploughman to steer and manoeuvre it.

When dragged through a field the coulter cuts down into the soil and the share cuts horizontally from the previous furrow to the vertical cut. This releases a rectangular strip of sod that is then lifted by the share and carried by the mouldboard up and over, so that the strip of sod (slice of the topsoil) that is being cut lifts and rolls over as the plough moves forward, dropping back to the ground upside down into the furrow and onto the turned soil from the previous run down the field. Each gap in the ground where the soil has been lifted and moved across (usually to the right) is called a *furrow*. The sod that has been lifted from it rests at about a 45 degree angle in the next-door furrow and lies up the back of the sod from the previous run.

In this way, a series of ploughing runs down a field leaves a row of sods that lie partly in the furrows and partly on the ground lifted earlier. Visually, across the rows, there is the land (unploughed part) on the left, a furrow (half the width of the removed strip of soil) and the removed strip almost upside-down lying on about half of the previous strip of inverted soil, and so on across the field. Each layer of soil and the gutter it came from forms the classic furrow.

The mouldboard plough greatly reduced the amount of time needed to prepare a field, and as a consequence, allowed a farmer to work a larger area of land. In addition, the resulting pattern of low (under the mouldboard) and high (beside it) ridges in the soil forms water channels, allowing the soil to drain. In areas where snow buildup causes difficulties, this lets farmers plant the soil earlier, as the snow runoff drains away more quickly.

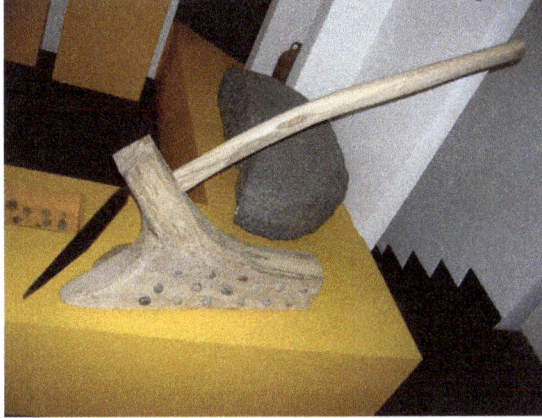

A reconstruction of a mouldboard plough

There are five major parts of a mouldboard plough:

1. Mouldboard

2. Share

3. Landside

4. Frog

5. Tailpiece

A *runner* extending from behind the share to the rear of the plough controls the direction of the plough, because it is held against the bottom land-side corner of the new furrow being formed. The holding force is the weight of the sod, as it is raised and rotated, on the curved surface of the mouldboard. Because of this runner, the mouldboard plough is harder to turn around than the scratch plough, and its introduction brought about a change in the shape of fields — from mostly square fields into longer rectangular "strips" (hence the introduction of the furlong).

An advance on the basic design was the *iron ploughshare*, a replaceable horizontal cutting surface mounted on the tip of the share. The earliest ploughs with a detachable and replaceable share date from around 1000 BC in the Ancient Near East, and the earliest iron ploughshares from ca. 500 BC in China. Early mouldboards were basically wedges that sat inside the cut formed by the coulter, turning over the soil to the side. The ploughshare spread the cut horizontally below the surface, so when the mouldboard lifted it, a wider area of soil was turned over. Mouldboards are known in Britain from the late 6th century on.

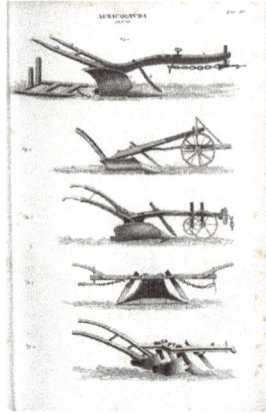

19th century editing plough

Loy Ploughing

Loy ploughing was a form of manual ploughing in Ireland, on very small farms — or on very hilly ground, where horses could not work or where farmers could not afford them. It was used up until the 1960s in poorer land. This suited the moist climate of Ireland, as the trenches formed by turning in the sods providing drainage. It also allowed the growing of potatoes in bogs as well as on mountain slopes where no other cultivation could take place.

Heavy Ploughs

Chinese iron plough with curved mouldboard, 1637

In the basic mouldboard plough the depth of the cut is adjusted by lifting against the runner in the furrow, which limited the weight of the plough to what the ploughman could easily lift. This limited the construction to a small amount of wood (although metal edges were possible). These ploughs were fairly fragile, and were not suitable for breaking up the heavier soils of northern Europe. The introduction of wheels to replace the runner allowed the weight of the plough to increase, and in turn allowed the use of a much larger mouldboard faced in metal. These *heavy ploughs* led to greater food production and eventually a significant population increase around AD 600.

Before the Han Dynasty (202 BC –AD 220), Chinese ploughs were made almost entirely of wood, except the iron blade of the ploughshare. By the Han period, the entire ploughshare was made of cast iron; these are the first known heavy mouldboard iron ploughs.

The Romans achieved the heavy wheeled mouldboard plough in the late 3rd and 4th century AD, when archaeological evidence appears, inter alia, in Roman Britain. The first indisputable appearance after the Roman period is from 643, in a northern Italian document. Old words connected with the heavy plough and its use appear in Slavic, suggesting possible early use in this region. The general adoption of the carruca heavy plough in Europe appears to have accompanied the adoption of the three-field system in the later eighth and early ninth centuries, leading to an improvement of the agricultural productivity per unit of land in northern Europe. This was accompanied by larger fields as well, known variously as carucates, ploughlands, and ploughgates.

Improved Designs

'A Champion ploughman', from Australia, circa 1900

The basic plough, with coulter, ploughshare and mouldboard remained in use for a millennium. Major changes in design did not become common until the Age of Enlightenment, when there was rapid progress in design. Joseph Foljambe in Rotherham, England, in 1730 used new shapes as the basis for the Rotherham plough, which also covered the mouldboard with iron. Unlike the heavy plough, the Rotherham (or Rotherham swing) plough consisted entirely of the coulter, mouldboard and handles. It was much lighter than conventional designs and became very popular in England. It may have been the first plough to be widely built in factories and the first to be commercially successful.

In 1789 Robert Ransome, an iron founder in Ipswich started casting ploughshares in a disused malting at St Margaret's Ditches. As a result of a mishap in his foundry, a broken mould caused molten metal to come into contact with cold metal, making the metal surface extremely hard — chilled casting — which he advertised as "self sharpening" ploughs, and received patents for his discovery.

James Small further advanced the design. Using mathematical methods he experimented with various designs until he arrived at a shape cast from a single piece of iron, an improvement on the *Scots plough* of James Anderson of Hermiston. A single-piece cast iron plough was also developed and patented by Charles Newbold in the United States. This was again improved on by Jethro Wood, a blacksmith of Scipio, New York, who made a three-part Scots Plough that allowed a broken piece to be replaced. In 1837 John Deere introduced the first steel plough; it was so much stronger than iron designs that it could work soil in areas of the US that had previously been considered unsuitable for farming.

Improvements on this followed developments in metallurgy: steel coulters and shares with softer iron mouldboards to prevent breakage, the chilled plough (an early example of surface-hardened steel), and eventually the face of the mouldboard grew strong enough to dispense with the coulter.

Single-sided Ploughing

Single-sided ploughing in a ploughing match

The first mouldboard ploughs could only turn the soil over in one direction (conventionally always to the right), as dictated by the shape of the mouldboard, and so the field had to be ploughed in long strips, or *lands*. The plough was usually worked clockwise around each land, ploughing the long sides and being dragged across the short sides without ploughing. The length of the strip was limited by the distance oxen (or later horses) could comfortably work without a rest, and their width by the distance the plough could conveniently be dragged. These distances determined the traditional size of the strips: a furlong, (or "furrow's length", 220 yards (200 m)) by a chain (22 yards (20 m)) — an area of one acre (about 0.4 hectares); this is the origin of the acre. The one-sided action gradually moved soil from the sides to the centre line of the strip. If the strip was in the same place each year, the soil built up into a ridge, creating the ridge and furrow topography still seen in some ancient fields.

Turnwrest Plough

The turnwrest plough allows ploughing to be done to either side. The mouldboard is removable, turning to the right for one furrow, then being moved to the other side of the plough to turn to the left (the coulter and ploughshare are fixed). In this way adjacent furrows can be ploughed in opposite directions, allowing ploughing to proceed continuously along the field and thus avoiding the ridge and furrow topography.

Reversible Plough

A four-furrow reversible Kverneland plough.

The reversible plough has two mouldboard ploughs mounted back-to-back, one turning to the right, the other to the left. While one is working the land, the other is carried upside-down in the air. At the end of each row, the paired ploughs are turned over, so the other can be used. This returns along the next furrow, again working the field in a consistent direction.

Riding and Multiple-furrow Ploughs

Early tractor-drawn two-furrow plough.

Early steel ploughs, like those for thousands of years prior, were *walking ploughs*, directed by the ploughman holding onto handles on either side of the plough. The steel ploughs were so much easier to draw through the soil that the constant adjustments of the blade to react to roots or clods was no longer necessary, as the plough could easily cut through them. Consequently, it was not long after that the first *riding ploughs* appeared. On these, wheels kept the plough at an adjustable level above the ground, while the ploughman sat on a seat; whereas, with earlier plows the plowman would have had to walk. Direction was now controlled mostly through the draught team, with levers allowing fine adjustments. This led very quickly to riding ploughs with multiple mould-boards, dramatically increasing ploughing performance.

A single draught horse can normally pull a single-furrow plough in clean light soil, but in heavier soils two horses are needed, one walking on the land and one in the furrow. For ploughs with two or more furrows more than two horses are needed and, usually, one or more horses have to walk on the loose ploughed sod—and that makes hard going for them, and the horse treads the newly ploughed land down. It is usual to rest such horses every half-hour for about ten minutes.

Heavy volcanic loam soils, such as are found in New Zealand, require the use of four heavy draught horses to pull a double-furrow plough. Where paddocks are more square than long-rectangular it is more economical to have horses four wide in harness than two-by-two ahead, thus one horse is always on the ploughed land (the sod). The limits of strength and endurance of horses made greater than two-furrow ploughs uneconomic to use on one farm.

Amish farmers tend to use a team of about seven horses or mules when spring plough-ing and as Amish farmers often help each other plough, teams are sometimes changed at noon. Using this method about 10 acres (4.0 ha) can be ploughed per day in light soils and about 2 acres (0.81 ha) in heavy soils.

Improvement in Metallurgy and Design

John Deere, a Vermont (U.S.A.) blacksmith, noted that the ploughing of many sticky, non-sandy soils might benefit from modifications in the design of the mouldboard and in the metals used. He noted that a polished needle would enter leather and fabric with greater ease, and a polished pitchfork required less effort as well. In the pursuit of a polished and thus slicker surface for a plough, he experimented with portions of saw blades and by 1837, he was making polished, cast steel ploughs. The energy effort required was lessened, which enabled the use of larger ploughs, making more effective use of horse power.

Balance Plough

The advent of the mobile steam engine allowed steam power to be applied to ploughing from about 1850. In Europe, soil conditions were often too soft to support the weight of heavy traction engines. Instead, counterbalanced, wheeled ploughs, known as *balance ploughs*, were drawn by cables across the fields by pairs of ploughing engines on opposite field edges, or by a single engine drawing directly towards it at one end and drawing away from it via a pulley at the other end. The balance plough had two sets of ploughs facing each other, arranged so when one was in the ground, the other set was lifted into the air. When pulled in one direction, the trailing ploughs were lowered onto the ground by the tension on the cable. When the plough reached the edge of the field, the other engine pulled the opposite cable, and the plough tilted (balanced), putting the other set of shares into the ground, and the plough worked back across the field.

A German balance plough. The left-turning set of shares have just completed a pass, and the right-turning shares are about to enter the ground to return across the field.

One set of ploughs was right-handed, and the other left-handed, allowing continuous ploughing along the field, as with the turnwrest and reversible ploughs. The man credited with the invention of the ploughing engine and the associated balance plough, in the mid nineteenth century, was John Fowler, an English agricultural engineer and inventor.

In America the firm soil of the Plains allowed direct pulling with steam tractors, such as the big Case, Reeves or Sawyer-Massey breaking engines. Gang ploughs of up to fourteen bottoms were used. Often these big ploughs were used in regiments of engines, so that in a single field there might be ten steam tractors each drawing a plough. In this way hundreds of acres could be turned over in a day. Only steam engines had the power to draw the big units. When internal combustion engines appeared, they had neither the strength nor the ruggedness compared to the big steam tractors. Only by reducing the number of shares could the work be completed.

Stump-jump Plough

Disc ploughs in Australia, circa 1900

The stump-jump plough was an Australian invention of the 1870s, designed to cope with the breaking up of new farming land, that contains many tree stumps and rocks that would be very expensive to remove. The plough uses a moveable weight to hold the ploughshare in position. When a tree stump or other obstruction such as a rock is encountered, the ploughshare is thrown upwards, clear of the obstacle, to avoid breaking the plough's harness or linkage; ploughing can be continued when the weight is returned to the earth after the obstacle is passed.

A simpler system, developed later, uses a concave disc (or a pair of them) set at a large angle to the direction of progress, that uses the concave shape to hold the disc into the soil — unless something hard strikes the circumference of the disk, causing it to roll up and over the obstruction. As the arrangement is dragged forward, the sharp edge of the disc cuts the soil, and the concave surface of the rotating disc lifts and throws the soil to the side. It doesn't make as good a job as the mouldboard plough (but this is not considered a disadvantage, because it helps fight wind erosion), but it does lift and break up the soil

Modern Ploughs

A British woman ploughing on a World War I recruitment poster for the Women's Land Army.

Modern ploughs are usually multiple reversible ploughs, mounted on a tractor via a three-point linkage. These commonly have between two and as many as seven mould-boards — and *semi-mounted* ploughs (the lifting of which is supplemented by a wheel about halfway along their length) can have as many as eighteen mouldboards. The hydraulic system of the tractor is used to lift and reverse the implement, as well as to adjust furrow width and depth. The ploughman still has to set the draughting linkage from the tractor so that the plough is carried at the proper angle in the soil. This angle and depth can be controlled automatically by modern tractors. As a complement to the rear plough a two or three mouldboards-plough can be mounted on the front of the tractor if it is equipped with front three-point linkage.

Specialist Ploughs

Chisel Plough

The *chisel plough* is a common tool to get deep tillage (prepared land) with limited soil disruption. The main function of this plough is to loosen and aerate the soils while leaving crop residue at the top of the soil. This plough can be used to reduce the effects of compaction and to help break up ploughpan and hardpan. Unlike many other ploughs the chisel will not invert or turn the soil. This characteristic has made it a useful addition to no-till and low-till farming practices that attempt to maximise the erosion-prevention benefits of keeping organic matter and farming residues present on the soil surface through the year. Because of these attributes, the use of a chisel plough is considered by some to be more sustainable than other types of plough, such as the mouldboard plough.

A modern John Deere 8110 Farm Tractor using a chisel plough. The ploughing tines are at the rear; the refuse-cutting coulters at the front.

Bigham Brother Tomato Tiller

The chisel plough is typically set to run up to a depth of eight to twelve inches (200 to 300 mm). However some models may run much deeper. Each of the individual ploughs, or shanks, are typically set from nine inches (229 mm) to twelve inches (305 mm) apart. Such a plough can encounter significant soil drag, consequently a tractor of sufficient power and good traction is required. When planning to plough with a chisel plough it is important to bear in mind that 10 to 20 horsepower (7.5 to 15 kW) per shank will be required, depending on depth.

Cultivators are often similar in form to chisel ploughs, but their goals are different. Cultivator teeth work near the surface, usually for weed control, whereas chisel plough shanks work deep beneath the surface. Consequently, cultivating also takes much less power per shank than does chisel ploughing.

Ridging Plough

A ridging plough is used for crops, such as potatoes or scallions, which are grown buried in ridges of soil using a technique called *ridging* or *hilling*. A ridging plough has two mouldboards facing away from each other, cutting a deep furrow on each pass, with high ridges either side. The same plough may be used to split the ridges to harvest the crop.

Scottish Hand Plough

This is a variety of ridge plough notable in that the blade points towards the operator. It is used solely by human effort rather than with animal or machine assistance, and is pulled backwards by the operator, requiring great physical effort. It is particularly used for second breaking of ground, and for potato planting. It is found in Shetland, some western crofts and more rarely Central Scotland. The tool is typically found on small holdings too small or poor to merit use of animals.

Mole Plough

The *mole plough* or *subsoiler* allows underdrainage to be installed without trenches, or it breaks up deep impermeable soil layers that impede drainage. It is a very deep plough, with a torpedo-shaped or wedge-shaped tip, and a narrow blade connecting this to the body. When dragged through the ground, it leaves a channel deep under the ground, and this acts as a drain. Modern mole ploughs may also bury a flexible perforated plastic drain pipe as they go, making a more permanent drain — or they may be used to lay pipes for water supply or other purposes. Similar machines, so called pipe-and-cable-laying ploughs, are even used under the sea, for the laying of cables, as well as preparing the earth for side-scan sonar in a process used in oil exploration.

Paraplough

The paraplough or paraplow is a tool for loosening compacted soil layers 12 to 16 inches deep and still maintain high surface residue levels.

Spade Plough

The spade plough is designed to cut the soil and turn it on its side, minimizing the damage to the earthworms, soil microorganism, and fungi. This helps maximize the sustainability and long term fertility of the soils.

Effects of Mouldboard Ploughing

Mouldboard ploughing, in cold and temperate climates, no deeper than 20 cm, aerates the soil by loosening it. It incorporates crop residues, solid manures, limestone and commercial fertilizers along with oxygen. By doing so, it reduces nitrogen losses by denitrification, accelerates mineralization and increases short-term nitrogen availability for transformation of organic matter into humus. It erases wheel tracks and ruts caused by harvesting equipment. It controls many perennial weeds and pushes back the growth of other weeds until the following spring. It accelerates soil warming and water evaporation in spring because of the lesser quantity of residues on the soil surface. It facilitates seeding with a lighter seeder. It controls many enemies of crops (slugs, crane flies, seedcorn maggots-bean seed flies, borers). It increases the number of "soil-eating" earthworms (endogea) but is detrimental to vertical-dwelling earthworms (anecic).

Ploughing leaves very little crop residue on the surface, which otherwise could reduce both wind and water erosion. Over-ploughing can lead to the formation of hardpan. Typically farmers break up hardpan up with a subsoiler, which acts as a long, sharp knife to slice through the hardened layer of soil deep below the surface. Soil erosion due to improper land and plough utilization is possible. Contour ploughing mitigates soil erosion by ploughing across a slope, along elevation lines. Alternatives to ploughing, such as the no till method, have the potential to actually build soil levels and humus. These may be suitable to smaller, more intensively cultivated plots, and to farming on poor, shallow or degraded soils that ploughing would further degrade.

- Plows in art

Back side of a 100 Mark banknote issued 1908

1975 Italian Lira coin

Broadcast Spreader

A broadcast seeder, alternately called a broadcast spreader or centrifugal fertilizer spreader(europe), is a farm implement commonly used for spreading seed, lime, fertilizer, sand, ice melt, etc., and is an alternative to drop spreaders/seeders.

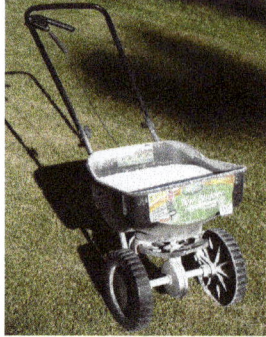

Hand-pushed broadcast spreader

Types

ATV tow spreaders normally have a larger capacity to enable the coverage of larger areas

The smallest are handheld with a hopper of several liters and which operate via hand cranking. A bit larger are push units with the spinning disk powered by gearing to the wheels. The next size up is designed to be towed behind a garden tractor or ATV. Very similar in size to the tow behind units are broadcast seeders that mount to the three-point hitch of a compact utility tractor, these are ideal for landscape and small property maintenance. Still larger are commercial broadcast seeders/spreaders designed and sized appropriately for agricultural tractors and mount to the tractor's three point hitch. The broadcast seeders that are mounted to a three point hitch are powered by a power take-off (P.T.O.) shaft from the tractor. At the largest size are pull behind or chassis mounted units for agricultural use that can spread widths of up to 90 feet.

How They Work

View of a tractor-operated broadcast seeder moved by three-point hitch and driven by PTO shaft

The basic operating concept of broadcast spreads is simple. A large material hopper is positioned over a horizontal spinning disk, the disk has a series of 3 or 4 fins attached to it which throw the dropped materials from the hopper out and away from the seeder/spreader. Alternately a pendulum spreading mechanism may be employed, this method is more common in mid-sized commercial spreaders for improved consistency in spreading. The photos clearly show the material hopper. Hoppers are commonly made of plastic, painted steel, or stainless steel. Stainless steel is usually used in large commercial units for strength and because granular fertilizer is often quite corrosive.

Some seeders/spreaders have directional fins to control the direction of the material that is thrown from the spreader. All broadcast spreaders require some form of power to spin the disk. On hand carried units, a hand crank spins gears to turn the disk. On tow behind units, the wheels spin a shaft that turns gears which, in turn, spin the disk. As is partially visible in one of the photos, with tractor mounted units, a mechanical P.T.O. shaft connected to the tractor and controlled by the tractor operator, spins the disk. There are some seeder/spreaders made for garden size tractors that use a 12 volt motor to spin the dispersing disk and yaw.

Planter (Farm Implement)

A two row planter featuring John Deere "71 Flexi" row units

Like a grain drill a planter is an agricultural farm implement towed behind a tractor, used for sowing crops through a field. It is connected to the tractor with a draw-bar, or a three-point hitch. Planters lay the seeds down in precise manner along rows. Seeds are distributed through devices called row units. The row units are spaced evenly along the planter. Planters vary greatly in size, from 1 row to 48, with the biggest in the world being the 48-row John Deere DB120. The space between the row units also vary greatly. The most common row spacing in the United States today is 30 inches.

On smaller and older planters, a marker extends out to the side half the width of the planter and creates a line in the field where the tractor should be centered for the next pass. The marker is usually a single disc harrow disc on a rod on each side of the planter. On larger and more modern planters, GPS navigation and auto-steer systems for the tractor are often used, eliminating the need for the marker. Some precision farming equipment such as Case IH AFS uses GPS/RKS and computer-controlled planter to sow seeds to precise position accurate within 2 cm. In an irregularly shaped field, the precision farming equipment will automatically hold the seed release over area already sewn when the tractor has to run overlapping pattern to avoid obstacles such as trees.

John Deere MaxEmerge XP Planter with Case IH AFS precision farming system which auto-steers using GPS

Older planters commonly have a seed bin for each row and a fertilizer bin for two or more rows. In each seed bin plates are installed with a certain number of teeth and tooth spacing according to the type of seed to be sown and the rate at which the seeds are to be sown. The tooth size (actually the size of the space between the teeth) is just big enough to allow one seed in at a time but not big enough for two. Modern planters often have a large bin for seeds that are distributed to each row known as central commodity systems.

Drive Systems

There are different types of planters available with the main difference being mechanical driven vs hydraulic/electrical driven. In a mechanical drive system the unit works by a small suspended tire being driven by another which is in contact with the ground (driven) tire. As the operator lowers the planter the two tires make contact and the planter is engaged. When the driven wheel begins to turn it then turns a series of gears that determine the population of the seed produced. The gears can be changed by the operator in order to change the planting population. A hydraulic driven system came about to correct the shortfalls of the ground driven system. Hydraulic driven systems allow the operator to change population on the go, as well as allowing the computer controller to follow a prepared prescription for a individual field. The system also allowed for plant populations to be infinite in that mechanical gears systems are limited to set number of population settings and gears available from manufactures. In 2014 John Deere introduced the ExactEmerge row unit which introduced high-speed planting. Precision Planting (later purchased by John Deere from Monsanto) followed suit and released the vDrive system. These system were unique, not that they electrical, but that they allowed an operator to double their speed when planting. Other manufactures had already developed an electrical planter, but lacked these additional improvements. Traditionally, an operator would plant at about 4.5-5.5 mph for optimal performance, however, with the advent of these systems electrical motors match the speed of the tractor and "dead-drop" the seed in the trench using either a belt or brush-belt which cause the forward momentum of the planter to be offset by the rearward momentum of the seed. Older systems would instead drop the seed through a tube after the meter rather than place is in the seed trench directly.

Drip Irrigation

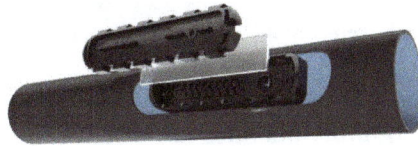

Open pressure compensated dripper

Drip irrigation is a form of irrigation that saves water and fertilizer by allowing water to drip slowly to the roots of many different plants, either onto the soil surface or directly onto the root zone, through a network of valves, pipes, tubing, and emitters. It is done through narrow tubes that deliver water directly to the base of the plant. It is chosen instead of surface irrigation for various reasons, often including concern about minimizing evaporation.

History

Drip irrigation in Mexico vineyard, 2000

Primitive drip irrigation has been used since ancient times. Fan Sheng-Chih Shu, written in China during the first century BCE, describes the use of buried, unglazed clay pots filled with water as a means of irrigation. Modern drip irrigation began its development in Germany in 1860 when researchers began experimenting with subsurface irrigation using clay pipe to create combination irrigation and drainage systems. Research was later expanded in the 1920s to include the application of perforated pipe systems. The usage of plastic to hold and distribute water in drip irrigation was later developed in Australia by Hannis Thill.

Usage of a plastic emitter in drip irrigation was developed in Israel by Simcha Blass and his son Yeshayahu. Instead of releasing water through tiny holes easily blocked by tiny particles, water was released through larger and longer passageways by using velocity to slow water inside a plastic emitter. The first experimental system of this type was established in 1959 by Blass who partnered later (1964) with Kibbutz Hatzerim to create an irrigation company called Netafim. Together they developed and patented the first practical surface drip irrigation emitter.

In the United States, the first drip tape, called *Dew Hose*, was developed by Richard Chapin of Chapin Watermatics in the early 1960s.

Modern drip irrigation has arguably become the world's most valued innovation in agriculture since the invention of the impact sprinkler in the 1930s, which offered the first practical alternative to surface irrigation. Drip irrigation may also use devices called micro-spray heads, which spray water in a small area, instead of dripping emitters. These are generally used on tree and vine crops with wider root zones. Subsurface drip irrigation (SDI) uses permanently or temporarily buried dripperline or drip tape located at or below the plant roots. It is becoming popular for row crop irrigation, especially in areas where water supplies are limited or recycled water is used for irrigation. Careful study of all the relevant factors like land topography, soil, water, crop and agro-climatic conditions are needed to determine the most suitable drip irrigation system and components to be used in a specific installation.

Components and Operation

Drip irrigation system layout and its parts

Water distribution in subsurface drip irrigation

Nursery flowers watered with drip irrigation in Israel

Components used in drip irrigation (listed in order from water source) include:

- Pump or pressurized water source

- Water filter(s) or filtration systems: sand separator, Fertigation systems (Venturi injector) and chemigation equipment (optional)

- Backwash controller (Backflow prevention device)

- Pressure Control Valve (pressure regulator)

- Distribution lines (main larger diameter pipe, maybe secondary smaller, pipe fittings)

- Hand-operated, electronic, or hydraulic control valves and safety valves

- Smaller diameter polytube (often referred to as "laterals")

- Poly fittings and accessories (to make connections)

- Emitting devices at plants (emitter or dripper, micro spray head, inline dripper or inline driptube)

In drip irrigation systems, pump and valves may be manually or automatically operated by a controller.

Most large drip irrigation systems employ some type of filter to prevent clogging of the small emitter flow path by small waterborne particles. New technologies are now being offered that minimize clogging. Some residential systems are installed without additional filters since potable water is already filtered at the water treatment plant. Virtually all drip irrigation equipment manufacturers recommend that filters be employed and generally will not honor warranties unless this is done. Last line filters just before the final delivery pipe are strongly recommended in addition to any other filtration system due to fine particle settlement and accidental insertion of particles in the intermediate lines.

Drip and subsurface drip irrigation is used almost exclusively when using recycled municipal waste water. Regulations typically do not permit spraying water through the air that has not been fully treated to potable water standards.

Because of the way the water is applied in a drip system, traditional surface applications of timed-release fertilizer are sometimes ineffective, so drip systems often mix liquid fertilizer with the irrigation water. This is called fertigation; fertigation and chemigation (application of pesticides and other chemicals to periodically clean out the system, such as chlorine or sulfuric acid) use chemical injectors such as diaphragm pumps, piston pumps, or aspirators. The chemicals may be added constantly whenever the system is irrigating or at intervals. Fertilizer savings of up to 95% are being reported from recent university field tests using drip fertigation and slow water delivery as compared to timed-release and irrigation by micro spray heads.

Properly designed, installed, and managed, drip irrigation may help achieve water conservation by reducing evaporation and deep drainage when compared to other types of irrigation such as flood or overhead sprinklers since water can be more precisely applied to the plant roots. In addition, drip can eliminate many diseases that are spread through water contact with the foliage. Finally, in regions where water supplies are severely limited, there may be no actual water savings, but rather simply an increase in production while using the same amount of water as before. In very arid regions or on sandy soils, the preferred method is to apply the irrigation water as slowly as possible.

Pulsed irrigation is sometimes used to decrease the amount of water delivered to the plant at any one time, thus reducing runoff or deep percolation. Pulsed systems are typically expensive and require extensive maintenance. Therefore, the latest efforts by emitter manufacturers are focused toward developing new technologies that deliver irrigation water at ultra-low flow rates, i.e. less than 1.0 liter per hour. Slow and even delivery further improves water use efficiency without incurring the expense and complexity of pulsed delivery equipment.

An emitting pipe is a type of drip irrigation tubing with emitters pre-installed at the factory with specific distance and flow per hour as per crop distance.

An emitter restricts water flow passage through it, thus creating head loss required (to the extent of atmospheric pressure) in order to emit water in the form of droplets. This head loss is achieved by friction / turbulence within the emitter.

Advantages and Disadvantages

Drip irrigation and spare drip irrigation tubes in banana farm at Chinawal, India

Pot irrigation by On-line drippers

The advantages of drip irrigation are:

- Fertilizer and nutrient loss is minimized due to localized application and reduced leaching.

- Water application efficiency is high if managed correctly

- Field levelling is not necessary.

- Fields with irregular shapes are easily accommodated.

- Recycled non-potable water can be safely used.

- Moisture within the root zone can be maintained at field capacity.

- Soil type plays less important role in frequency of irrigation.

- Soil erosion is lessened.

- Weed growth is lessened.

- Water distribution is highly uniform, controlled by output of each nozzle.

- Labour cost is less than other irrigation methods.

- Variation in supply can be regulated by regulating the valves and drippers.

- Fertigation can easily be included with minimal waste of fertilizers.

- Foliage remains dry, reducing the risk of disease.

- Usually operated at lower pressure than other types of pressurised irrigation, reducing energy costs.

The disadvantages of drip irrigation are:

- Initial cost can be more than overhead systems.

- The sun can affect the tubes used for drip irrigation, shortening their usable life.

(This section does not include a discussion of the effects of degrading plastic on the soil content and subsequent effect on food crops. With many types of plastic, when the sun degrades the plastic, causing it to become brittle, the estrogenic chemicals (that is, chemicals replicating female hormones) which would cause the plastic to retain flexibility have been released into the surrounding environment.)

- If the water is not properly filtered and the equipment not properly maintained, it can result in clogging.

- For subsurface drip the irrigator cannot see the water that is applied. This may lead to the farmer either applying too much water (low efficiency) or an insufficient amount of water, this is particularly common for those with less experience with drip irrigation.

- Drip irrigation might be unsatisfactory if herbicides or top dressed fertilizers need sprinkler irrigation for activation.

- Drip tape causes extra cleanup costs after harvest. Users need to plan for drip tape winding, disposal, recycling or reuse.

- Waste of water, time and harvest, if not installed properly. These systems require careful study of all the relevant factors like land topography, soil, water, crop and agro-climatic conditions, and suitability of drip irrigation system and its components.

- In lighter soils subsurface drip may be unable to wet the soil surface for germination. Requires careful consideration of the installation depth.

- most drip systems are designed for high efficiency, meaning little or no leaching fraction. Without sufficient leaching, salts applied with the irrigation water may build up in the root zone, usually at the edge of the wetting pattern. On the other hand, drip irrigation avoids the high capillary potential of traditional surface-applied irrigation, which can draw salt deposits up from deposits below.

- the PVC pipes often suffer from rodent damage, requiring replacement of the entire tube and increasing expenses.

- Drip irrigation systems cannot be used for damage control by night frosts (like in the case of sprinkler irrigation systems)

Uses

Drip irrigation is used in farms, commercial greenhouses, and residential gardeners. Drip irrigation is adopted extensively in areas of acute water scarcity and especially for crops and trees such as coconuts, containerized landscape trees, grapes, bananas, pandey, eggplant, citrus, strawberries, sugarcane, cotton, maize, and potatoes.

Irrigation dripper

Drip irrigation for garden available in drip kits are increasingly popular for the home-owner and consist of a timer, hose and emitter. Hoses that are 4 mm in diameter are used to irrigate flower pots.

Grain Drying

Grain drying is process of drying grain to prevent spoilage during storage. The grain drying described in this article is that relating to processes supplementary to the natural ones.

Overview

Continuous mixed flow grain dryer

Hundreds of millions of tonnes of wheat, corn, soybean, rice and other grains as sorghum, sunflower seeds, rapeseed/canola, barley, oats, etc., are dried in grain dryers. In the main agricultural countries, drying comprises the reduction of moisture from about 17-30%w/w to values between 8 and 15%w/w, depending on the grain. The final moisture content for drying must be adequate for storage. The more oil the grain has, the lower its storage moisture content will be (though its initial moisture for drying will also be lower). Cereals are often dried to 14% w/w, while oilseeds, to 12.5% (soybeans),

8% (sunflower) and 9% (peanuts). Drying is carried out as a requisite for safe storage, in order to inhibit microbial growth. However, low temperatures in storage are also highly recommended to avoid degradative reactions and, especially, the growth of insects and mites. A good maximum storage temperature is about 18 °C.

The largest dryers are normally used "Off-farm", in elevators, and are of the continuous type: Mixed-flow dryers are preferred in Europe, while Cross-flow dryers in the United States. In Argentina, both types are commonly found. Continuous flow dryers may produce up to 100 metric tonnes of dried grain per hour. The depth of grain the air must traverse in continuous dryers range from some 0.15 m in Mixed flow dryers to some 0.30 m in Cross-Flow. Batch dryers are mainly used "On-Farm", particularly in the United States and Europe. They normally consist of a bin, with heated air flowing horizontally from an internal cylinder through an inner perforated metal sheet, then through an annular grain bed, some 0.50 m thick (coaxial with the internal cylinder) in radial direction, and finally across the outer perforated metal sheet, before being discharged to the atmosphere. The usual drying times range from 1 h to 4 h depending on how much water must be removed, type of grain, air temperature and the grain depth. In the United States, continuous counterflow dryers may be found on-farm, adapting a bin to slowly drying grain fed at the top and removed at the bottom of the bin by a sweeping auger.

Grain drying is an active area of manufacturing and research. The performance of a dryer can be simulated with computer programs based on mathematical models that represent the phenomena involved in drying: physics, physical chemistry, thermodynamics and heat and mass transfer. Most recently computer models have been used to predict product quality by achieving a compromise between drying rate, energy consumption, and grain quality. A typical quality parameter in wheat drying is breadmaking quality and germination percentage whose reductions in drying are somewhat related.

Grain Drying Fundamentals

Drying starts at the bottom of the bin, which is the first place air contacts. The dry air is brought up by the fan through a layer of wet grain. Drying happens in a layer of 1 to 2 feet thick, which is called the drying zone. The drying zone moves from the bottom of the bin to the top, and when it reaches the highest layer, the grain is dry. The grain below drying zone is in equilibrium moisture content with drying air, which means it is safe for storage; while the grain above still needs drying. The air is then forced out the bin through exhaust vent.

Allowable Storage Time

Allowable storage time is an estimate of how long the grain needs to be dried before spoilage and maintain grain quality during storage. In grain storage process, fungi or molds are the primary concern. Many other factors, such as insects, rodents, and bacteria, also affect the condition of storage. The lower the grain temperature is, the longer the allowable storage time will be.

Proper Moisture Levels for Safe Storage

It is possible for long period safe storage if grain moisture content is less than 14%, and stored away from insects, rodents and birds. The following figure is the recommended moisture content for safe storage.

Storage duration	Required MC for safe storage	Potential problems
Weeks to a few months storage	14% or less	Molds, discoloration, respiration loss, insect damage, moisture adsorption
Storage for 8 to 12 months	13% or less	Insect damage
Storage of farmer's seeds	12% or less	Loss of germination
Storage for more than 1 year	9% or less	Loss of germination

Equilibrium Moisture Content

Moisture content in grain is related to the relative humidity and the temperature of the surrounding air. Equilibrium moisture content point is the point when grain no longer losing or gaining water when contacting with drying air. The final moisture content of the grain is up to the amount of moisture in the drying air, which is the relative humidity. The low relative humidity means air is dry and it has a large potential of picking up water. The lower the relative humidity is, the drier the air is. In general, one-half reduce in relative humidity is caused by 20° degree increase in air temperature.

Temperature

Heated air may be used in grain drying process. It can not only accelerate moisture migration inside the kernel, but also can evaporate the moisture on the surface. The major problem about heated air for drying process is the kernel temperature. The grain kernel may be damaged by high kernel temperatures. Usually, kernel temperature is lower than the air temperature. For different use of the corn, temperatures vary. For example, for seed corn, the maximum temperature is 110 °F; for livestock feeding corn, the maximum temperature is 180 °F.

Aeration

Aeration process refers to the process of moving air through grain. Airflow is a measurement of the amount of air in cubic feet per minute (CFM). In grain drying process, drying time is largely depended on aeration rates. Without sufficient airflow, grain may be damaged before drying is complete. Fans are used to move air through grain.

Application	Grain aeration rate (cfm/bu)
Quality maintenance	1/50 to 1
Natural-air bin drying	1 to 3
Heated-air bin drying	2 to 12
Batch or continuous-flow column dryers	50 to 150
Fluidization	~400

Drying Cost

The drying cost is made up of two parts: the capital cost and the operating cost. Capital cost is largely depend on the drying rate requirement, and equipment cost. Operating cost refers to fuel, electricity and labor force cost. The amount of energy required to dry a bushel of grain is similar for all the drying methods. Some methods depend largely on natural air, while others may use LP heat or natural gas, which make energy cost vary. Basically, fuel and electrical power are the major portions of the operating cost. Drying cost is based on the B.T.U. consumption of temperature change from environment to desired one.

Classification of Grain Drying Methods

In-storage drying methods

1.   Low-temperature drying

2.   Multiple-layer drying

Batch drying methods

1.   Bin batch drying

2.   Column batch drying

Continuous flow drying methods

1.   Cross flow drying

2.   Counter flow drying

3.   Concurrent flow drying

In-storage Drying Methods

Low-temperature Drying

In-storage drying methods refers to those grain is dried and stored in the same container. Low-temperature drying, also known as near-ambient drying, is one of in-storage drying methods. There are four major factors which influence low temperature drying: the variability of weather, the harvest moisture content, the air flow in the storage bin and the amount of heated air. Most low-temperature dryers are built to dry grain as slowly as possible, while in the same time less spoilage on the grain. It is suggested that low temperature drying system is better operated when the average daily temperature is between 30 °C and 50 °C. Rather than control the drying air temperature, the low-temperature drying focuses on the relative humidity in order to achieve equilibrium moisture content (EMC) in all grain layers. Low-temperature drying process usually takes 5 days to several months depends on several important variables: weather, airflow, initial moisture content and amount of heat used. Among which, airflow is the key factor. Without appropriate airflow rate, spoilage will occur before drying is completed. By using heated air (LP heat, electric heat and solar heat), the relative humidity of the drying air is better controlled to achieve the desired moisture content. Usually, heated air dryer is used when the relative humidity larger than 70%. In electric heat dryers, an electrical resistant heater is usually placed before the fan to heat the airstream. In some case, a humidistat is employed to control the heater. In solar heat dryers, the drying air passes through the solar collector first to be heated (usually 10 to 12 °F rise), then enters the bin through the fan and motor. The advantages of in-storage low temperature drying are quick filling, high quality product, less equipment requirement; while the disadvantages are long drying time, electrical demand if using electric heat, high management skills and uncertain harvest moisture content.

Multiple-layer Drying

Multiple layer drying method refers to the use of LP heat or natural gas in drying corn. Compared to low-temperature methods, multiple-layer drying requires higher temperatures, which results in a shorter allowable storage time. Multiple-layer drying without stirring is the basic multiple-layer drying method, in which airstream is entered through an LP heater by a fan. Usually, the temperature rise after the LP burner is remained low in order to avoid over drying in the bottom layers in the bin. As soon as corn is dried in the bin, the burner is turned off and the fan is used to bring the corn to ambient temperature. The advantages of multiple-layer drying without stirring are little handling of corn, and bin can be used as either dryer or storage; the disadvantages are slow filling and over drying in the bottom layers (Bern and Brumm, 2010). Multiple-layer drying with stirring can not only dry grain equilibrium from top to bottom, but also decrease the air resistance of the grain. Moreover, using stirring system can avoid over drying in bottom layer problem and give a uniform grain moisture content in the whole bin. When drying is complete, the burner is turned off while the fan and stirrer are used to mix the corn to achieve equal moisture content and temperature. The advantages of adding stirring are preventing over drying and accelerating drying and allowable fill rate; the disadvantages of stirring system are additional expenses and decreasing bin capacity.

Batch Drying Methods

Bin-Batch Drying

In batch drying methods, certain amount of grain is placed first, usually 2 to 4 inches, the batch is dried and cooled later, then drying is stopped and batch is removed. The batch dryers are usually operating under this sequence and repeating this sequence for several times. The bin-batch drying methods employ a full perforated floor as the dryer. Without stirring, large variety of equipment is available and the batch can be used as both dryer and cooler, but there may be large moisture gradient from top to bottom and losing time in loading and unloading process. When adding stirring system, unequilibrium moisture content problem is avoided, however, stirrer is an added expense. When using bin-batch roof dryer, time losing problem can be solved. There is a drying floor under the bin roof and the drying fan and burner is installed high on the bin wall. When the drying process is completed, grain is put in the regular bin floor, thus unloading time is reduced. However, there is no wet grain holding in bin-batch roof dryers and there is more expense on machines.

Column Batch Drying

The column formed in this kind of dryer is made up of two vertical perforated steel sheets, which is about 12 inches thick each. The capacity of column batch dryers is too small to store grains. The advantages of column batch, stationary bed dryer are easy to move and the dyer can be used as cooler; while the disadvantages are time losing when cooling, loading and unloading and unequal moisture distribution when drying is completed. When column batch recirculating dryer is used, the moisture content variation problem is avoided, but the additional handling process may result in grain spoilage.

Continuous Flow Drying Methods

Cross Flow Drying

Cross flow dryers is one of the most widely used continuous flow dryers. In the cross flow dryer, the airstream is perpendicular to the grain flow. Then the grain near the drying air is over dried, while on the other side, grain is under dried. Moisture gradient exists when drying is complete. In reality, the lower the airflow rate, the higher the grain moisture content variation between two sides of the column.

Concurrent Flow Drying

In the concurrent flow dryer, both the grain and air are moving in the same direction, which means the wettest grain is subjected to the hottest drying air. The kernels leave the drying region at the same temperature and the same moisture content. Energy efficiency is 40% better compared to cross flow dryer. However, the bed depth must be deeper than 12 inches depth than cross flow type. Thus, fan power requirements are high in this type of dryer.

Counter Flow Drying

In the counter flow dryer, the grain and the air are moving in opposite directions, which mean the driest grain is subjected to the hottest drying air. The kernels leave the drying region at the same temperature and the same moisture as in concurrent flow dryers. The suggesting air temperatures are less than 180 °F because the driest kernels are more likely to be damaged by hot air.

Applications of Grain Drying

Sunflower Drying

For different types of sunflowers, the preservation moisture content is different. Oilseed sunflowers are better dried to 9 percent moisture content, while bird seed sunflowers are 10 percent moisture content. Compared to corn drying, sunflowers are more easily to be dried and kept in safe storage. What's more, high temperature may not have adverse effect on sunflowers kernel, which may be the reason of the fatty acid composition. There was no evidence of damage when air was heated up to 220 °F in drying. However, fine hairs and fibers on the seed coat of sunflowers may cause fire hazard. It is suggested that remove the flaming particles first when heating the sunflowers.

Bean Drying

The seed coat of bean is quite fragile and easy to damaged in cracking and splitting, which may cause loss to the producer. Some studies on beans suggested that in order to avoid cracking, it is better to keep drying air above 40 percent relative humidity.

Corn Drying

When drying corn kernels it is important to keep in mind that cracks and fractures in the corn can lead to many problems in both the storage and processing. The major problem that occurs from high temperature drying and then rapid cooling of the grain is stress-cracking. Stress-cracking is when fractures become present in the corn endosperm. Stress-cracked kernels often absorb water too quickly, are more likely to become broken, and are increasingly susceptible to insect and mold damage during dry storage. In order to reduce the amount of grain that is lost due to stress-cracking, medium temperature and slow cooling, or natural air and low temperature drying methods should be employed.

Mower

A mower is a person or machine that cuts (mows) grass or other plants that grow on the ground. Usually mowing is distinguished from reaping, which uses similar implements, but is the traditional term for harvesting grain crops, e.g. with reapers and combines.

A smaller mower used for lawns and sports grounds (playing fields) is called a *lawn mower* or *grounds mower*, which is often self-powered, or may also be small enough to be pushed by the operator. Grounds mowers have reel or rotary cutters. Larger mowers or *mower-conditioners* are mainly used to cut grass (or other crops) for hay or silage and often place the cut material into rows, which are referred to as *windrows*. *Swathers* (or *windrowers*) are also used to cut grass (and grain crops). Prior to the invention and adoption of mechanized mowers, (and today in places where use of a mower is impractical or uneconomical), grass and grain crops were cut by hand using scythes or sickles.

Mower Configurations

Larger mowers are usually *ganged* (equipped with a number or gang of similar cutting units), so they can adapt individually to ground contours. They may be powered and drawn by a tractor or draft animals. The cutting units can be mounted underneath the tractor between the front and rear wheels, mounted on the back with a three-point hitch or pulled behind the tractor as a trailer. There are also dedicated self-propelled cutting machines, which often have the mower units mounted at the front and sides for easy visibility by the driver. *Boom* or *side-arm* mowers are mounted on long hydraulic arms, similar to a backhoe arm, which allows the tractor to mow steep banks or around objects while remaining on a safer surface.

Mower Types

The cutting mechanism in a mower may be one of several different designs:

Sickle Mower

Eicher tractor with a mid-mounted finger-bar mower

Sickle mowers, also called *reciprocating mowers*, *bar mowers*, *sickle-bar mowers*, or *finger-bar mowers*, have a long (typically six to seven and a half feet) bar on which are mounted fingers with stationary guardplates. In a channel on the bar there is a reciprocating sickle with very sharp sickle sections (triangular blades). The sickle bar is driven back and forth along the channel. The grass, or other plant matter, is cut between the sharp edges of the sickle sections and the finger-plates (this action can be likened to an electric hair clipper).

The bar rides on the ground, supported on a skid at the inner end, and it can be tilted to adjust the height of the cut. A spring-loaded board at the outer end of the bar guides the cut hay away from the uncut hay. The so-formed channel, between cut and uncut material, allows the mower skid to ride in the channel and cut only uncut grass cleanly on the next swath. These were the first successful horse-drawn mowers on farms and the general principles still guide the design of modern mowers.

Rotary Mower

Rotary cutters mounted on a swather

Rotary mowers, also called *drum mowers*, have a rapidly rotating bar, or disks mounted on a bar, with sharpened edges that cut the crop. When these mowers are tractor-mounted they are easily capable of mowing grass at up to 20 miles per hour (32 km/h) in good conditions. Some models are designed to be mounted in double and triple sets on a tractor, one in the front and one at each side, thus able to cut up to 20 foot (6 metre) swaths.

In rough cutting conditions, the blades attached to the disks are swivelled to absorb blows from obstructions. Mostly these are rear-mounted units and in some countries are called *scrub cutters*. Self-powered mowers of this type are used for rougher grass in gardening and other land maintenance.

Reel Mower

Reel mower

Reel mowers, also called *cylinder mowers* (familiar as the hand-pushed or self-powered cylinder lawn mower), have a horizontally rotating cylindrical reel composed of helical blades, each of which in turn runs past a horizontal cutter-bar, producing a continuous scissor action. The bar is held at an adjustable level just above the ground and the reel runs at a speed dependent on the forward movement speed of the machine, driven by wheels running on the ground (or in self-powered applications by a motor). The cut grass may be gathered in a collection bin.

This type of mower is used to produce consistently short and even grass on bowling greens, lawns, parks and sports grounds. When pulled by a tractor (or formerly by a horse), these mowers are often ganged into sets of three, five or more, to form a *gang mower*. A well-designed reel mower can cut quite tangled and thick tall grass, but this type works best on fairly short, upright vegetation, as taller vegetation tends to be rolled flat rather than cut.

Home reel mowers have certain benefits over motor-powered mowers as they are quieter and not dependent on any extra form of power besides the person doing the mowing. This is useful not only to lessen dependence on other types of power which may have availability issues, but also lessens the impact on the environment.

Flail Mower

Flail mower

Flail mowers have a number of small blades on the end of chains attached to a horizontal axis. The cutting is carried out by the ax-like heads striking the grass at speed. These types are used on rough ground, where the blades may frequently be fouled by other objects, or on tougher vegetation than grass, such as brush (scrub). Due to the length of the chains and the higher weight of the blades, they are better at cutting thick brush than other mowers, because of the relatively high inertia of the blades. In some types the cut material may be gathered in a collection bin. As a boom mower (see above), a flail mower may be used in an upright position for trimming the sides of hedges, when it is often called a *hedge-cutter*.

Forestry Mulching

Fecon mulching attachment on a Sennebogen excavator, being used to clear roadside brush in Germany

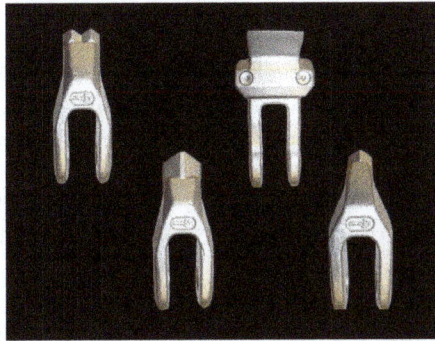

Examples of chipper tools ("teeth") available on forestry mulching attachments

Forestry mulching is a land clearing method that uses a single machine to cut, grind, and clear vegetation.

A forestry mulching machine, also referred to as a forestry mulcher, masticator, or brushcutter, uses a rotary drum equipped with steel chipper tools ("teeth") to shred vegetation. They are manufactured as application-specific tractors and as mulching attachments ("mulching heads") for existing tracked and rubber-tired forestry tractors, skid steers, or excavators.

Heavy duty forestry mulchers can clear up to fifteen acres of vegetation a day depending on terrain, density, and type of material. Forestry mulchers are often used for land clearing, right-of-way, pipeline/power line, and wildfire prevention and management, vegetation management, invasive species control, and wildlife restoration.

Applications of Forestry Mulching

Right-of-way Clearing and Maintenance

Forestry mulching is used in the right-of-way clearing and maintenance for roads, highways, pipelines, and other utility lines. This process often requires complete removal of standing trees, stumps, and vegetation.

Land Clearing

Forestry mulchers can be used in commercial and residential land clearing projects such as site preparation and development, cutting and clearing brush, nature and recreational trail creation, and seismic exploration.

Wetlands and Riparian Habitat Conservation

Forestry mulching has become popular among nonprofit riparian conservation organizations, government agencies, hunt clubs, and private land owners in attempts to maintain habitats for pheasants, doves, elk, deer and various other animals. Maintaining an animal habitat encompasses several different aspects: food, water, shelter, and space, and there are many products that can help reclaim and maintain wildlife habitats for these animals.

- Food: Forestry mulchers and forestry mowers are often used for removing underbrush and invasive species, such as buckthorn and multiflora rose, in order to allow the rejuvenation of grasses and other food sources.

- Water: Forestry mulchers and tree shears can be used to restore water source access that has been blocked due to tree and understory growth, allowing animals to access the water.

- Shelter and space: Mulching can remove invasive underbrush that prevents the growth of the grasses required by certain animals for shelter, breeding, escaping the summer heat, and protecting themselves from cold temperatures.

Invasive Species Control

Some common invasive plant species such as tamarisk (salt cedar), Pinyon-juniper (pj), Russian olive, buckthorn, and multiflora rose can invade a natural habitat, soak up a tremendous amount of ground water, and need to be removed to reestablish the native habitat or to preserve the water table. Invasive insects such as pine beetles can also devastate forests, leaving behind rotting trees with diminishing timber value and that may become falling hazards if they lose their ability to stand up against wind.

Proactive mulching can reduce stress on trees caused by crowding, making them less susceptible to attack from invasive species. Mulching invasive species in place can control the spread of invasive plants, insects, and fungus. The mulching action tends to discharge the material downward and within a reasonably confined area, versus other methods such as rotary cutters that may laterally disperse pine beetles or other invasive species into neighboring healthy trees.

Wildfire Prevention and Management

- Proactive mulching: Mulching reduces the potential for wildfires by eliminating small leafy plants, fallen or rotten trees, and other fuel sources. If left untreated, these fuel loads increase potential for fire, increase the heat intensity, and serve as fire ladders that enable fire to elevate quickly to the tops of trees which is where a fire can spread most quickly. Mulching can also be used to create a coarse grind finish that can create a more ideal controlled burn.

- Reactive mulching: In addition to proactive thinning of vegetation to mitigate fire fuels, forestry mulching can be used for reactive cutting of lines (fire breaks) on active fires. Larger forestry mulchers leave minimal cleanup requirements and can help reduce the overall costs of active fire mitigation.

- Cleanup: After the fire is out and cleanup efforts are under way, tracked forestry mulching machines, mulching attachments, and an excavator with a mulching attachment can provide a top layer of mulch to prevent soil erosion on slopes and minimize water pollution.

Advantages of Forestry Mulching

By processing trees and other vegetation where they stand, mulching machines eliminate many of the steps involved in land clearing such as site prep, cutting/felling/hauling, and site cleanup. This also eliminates the need for multiple machines such as a bulldozer accompanied by some combination of excavators, tree shears, wood chippers or grinders, and hauling equipment. One simpler jobs only one mulching machine is required, reducing fuel requirements and emissions.

Some mulching machines also have the ability to operate on steep slopes and in small or tight areas, in poor ground conditions, and in wet or snowy weather.

Mulching machines are capable of clearing land of unwanted trees and brush with limited disturbance to soils or desirable vegetation.

Traditional land clearing methods often present an increased risk of erosion by pushing over trees, uprooting the stump and roots, and substantially disturbing soils. In contrast, mulching the vegetation leaves the soil structure intact. The mulched material can be left on the ground and will act as an erosion barrier while returning nutrients back into the soil through decomposition. Over time, grass will naturally grow through the mulch and can be maintained with mowing.

Disadvantages of Forestry Mulching

Depending on the size and orientation of the mulching head, forestry mulchers can only fell smaller trees. Mulchers with a mulching head that rotates about a vertical axis can typically handle trees up to 6-8 inches in diameter, while mulchers with a mulching head that rotates about a horizontal axis can handle trees up to 30 inches in diameter. Mulching trees at the upper end of this size range can be dangerous for both the equipment and the operator. Mulching areas with a variety of vegetation and terrain may require multiple pieces of equipment, including tracked mulching machines, excavator mulchers, and skid steer tractors equipped with mulching attachments.

Rocks and stones cannot be processed or moved by the machines, and the teeth grinding against rocky ground can both wear them out and cause a fire hazard. Smaller rocks and other debris can be thrown through the air, and while these are usually deflected by a protective shroud, they can present a danger to the operator and surrounding people and structures.

Even for the largest machines, mulching is only effective when less than 25 tons of vegetation or 100 trees are present per acre.

Although mulching is significantly faster and less labor-intensive than land clearing by hand, it requires the site to have road access for fueling and maintenance.

Threshing Machine

A threshing machine in operation

The thrashing machine, or, in modern spelling, threshing machine (or simply thresher), was first invented by Scottish mechanical engineer Andrew Meikle for use in agriculture. It was devised (c. 1786) for the separation of grain from stalks and husks. For thousands of years, grain was separated by hand with flails, and was very laborious and time-consuming, taking about one-quarter of agricultural labor by the 18th century. Mechanization of this process took much of the drudgery out of farm labour.

Early Social Impacts

The Swing Riots in the UK were partly a result of the threshing machine. Following years of war, high taxes and low wages, farm labourers finally revolted in 1830. These farm labourers had faced unemployment for a number of years due to the widespread introduction of the threshing machine and the policy of enclosing fields. No longer were thousands of men needed to tend the crops, a few would suffice. With fewer jobs, lower wages and no prospects of things improving for these workers the threshing machine was the final straw, the machine was to place them on the brink of starvation. The Swing Rioters smashed threshing machines and threatened farmers who had them.

The riots were dealt with very harshly. Nine of the rioters were hanged and a further 450 were transported to Australia.

Later Adoption

Early threshing machines were hand-fed and horse-powered. They were small by today's standards and were about the size of an upright piano. Later machines were steam-powered, driven by a portable engine or traction engine. Isaiah Jennings, a skilled inventor, created a small thresher that doesn't harm the straw in the process. In 1834, John Avery and Hiram Abial Pitts devised significant improvements to a machine that automatically threshes and separates grain from chaff, freeing farmers from a slow and laborious process. Avery and Pitts were granted United States patent #542 on December 29, 1837.

John Ridley, an Australian inventor, also developed a threshing machine in South Australia in 1843.

The 1881 *Household Cyclopedia* said of Meikle's machine:

> "Since the invention of this machine, Mr. Meikle and others have progressively introduced a variety of improvements, all tending to simplify the labour, and to augment the quantity of the work performed. When first erected, though the grain was equally well separated from the straw, yet as the whole of the straw, chaff, and grain, was indiscriminately thrown into a confused heap, the work could only with propriety be considered as half executed. By the addition of rakes, or shakers, and two pairs of fanners, all driven by the same machinery, the different processes of thrashing, shaking, and winnowing are now all at once performed, and the grain immediately prepared for the public market. When it is added, that the quantity of grain gained from the superior powers of the machine is fully equal to a twentieth part of the crop, and that, in some cases, the expense of thrashing and cleaning the grain is considerably less than what was formerly paid for cleaning it alone, the immense saving arising from the invention will at once be seen.

> "The expense of horse labour, from the increased value of the animal and the charge of his keeping, being an object of great importance, it is recommended that, upon all sizable farms, that is to say, where two hundred acres [800,000 m²], or upwards, of grain are sown, the machine should be worked by wind, unless where local circumstances afford the conveniency of water. Where coals are plenty and cheap, steam may be advantageously used for working the machine."

Steam-powered machines used belts connected to a traction engine; often both engine and thresher belonged to a contractor who toured the farms of a district. Steam remained a viable commercial option until the early post-WWII years.

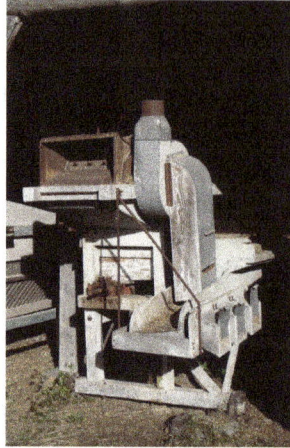

Open-air museum in Saint-Hubert, Belgium.

Farming Process

Threshing is just one step of the process in getting cereals to the grinding mill and customer. The wheat needs to be grown, cut, stooked (shocked, bundled), hauled, threshed, de-chaffed, straw baled, and then the grain hauled to a grain elevator. For many years each of these steps was an individual process, requiring teams of workers and many machines. In the steep hill wheat country of Palouse in the Northwest of the United States, steep ground meant moving machinery around was problematic and prone to rolling. To reduce the amount of work on the sidehills, the idea arose of combining the wheat binder and thresher into one machine, known as a combine harvester. About 1910, horse pulled combines appeared and became a success. Later, gas and diesel engines appeared with other refinements and specifications.

Modern Developments

In Europe and Americas

Threshing of paddy by machine, Bangladesh.

Modern day combine harvesters (or simply combines) operate on the same principles and use the same components as the original threshing machines built in the 19th century. Combines also perform the reaping operation at the same time. The name *combine* is derived from the fact that the two steps are combined in a single machine. Also, most modern combines are self-powered (usually by a diesel engine) and self-propelled, although tractor powered, pull type combines models were offered by John Deere and Case International into the 1990s.

Today, as in the 19th century, the threshing begins with a cylinder and concave. The cylinder has sharp serrated bars, and rotates at high speed (about 500 RPM), so that the bars beat against the grain. The concave is curved to match the curve of the cylinder, and serves to hold the grain as it is beaten. The beating releases the grain from the straw and chaff.

Whilst the majority of the grain falls through the concave, the straw is carried by a set of "walkers" to the rear of the machine, allowing any grain and chaff still in the straw to fall below. Below the straw walkers, a fan blows a stream of air across the grain, removing dust and fines and blowing them away.

The grain, either coming through the concave or the walkers, meets a set of sieves mounted on an assembly called a shoe, which is shaken mechanically. The top sieve has larger openings, and serves to remove large pieces of chaff from the grain. The lower sieve separates clean grain, which falls through, from incompletely threshed pieces. The incompletely threshed grain is returned to the cylinder by means of a system of conveyors, where the process repeats.

Some threshing machines were equipped with a bagger, which invariably held two bags, one being filled, and the other being replaced with an empty. A worker called a *sewer* removed and replaced the bags, and sewed full bags shut with a needle and thread. Other threshing machines would discharge grain from a conveyor, for bagging by hand. Combines are equipped with a grain tank, which accumulates grain for deposit in a truck or wagon.

A large amount of chaff and straw would accumulate around a threshing machine, and several innovations, such as the air chaffer, were developed to deal with this. Combines generally chop and disperse straw as they move through the field, though the chopping is disabled when the straw is to be baled, and chaff collectors are sometimes used to prevent the dispersal of weed seed throughout a field.

The corn sheller was almost identical in design, with slight modifications to deal with the larger kernel size and presence of cobs. Modern-day combines can be adjusted to work with any grain crop, and many unusual seed crops.

Both the older and modern machines require a good deal of effort to operate. The concave clearance, cylinder speed, fan velocity, sieve sizes, and feeding rate must be adjusted for crop conditions.

Another Development in Asia

Video of a petrol-powered machine threshing rice in Hainan, China

From the early 20th century, petrol or diesel-powered threshing machines, designed especially to thresh rice, the most important crop in Asia, have been developed along different lines to the modern combine.

Even after the combine was invented and became popular, a new compact-size thresher called a *harvester*, with wheels, still remains in use and at present it is available from a Japanese agricultural manufacturer. The compact-size machine is very convenient to handle in small terrace fields in mountain areas where a large machine, such as combine, is not usable.

People there use this harvester with a modern compact binder.

Preservation

A number of older threshing machines have survived into preservation. They are often to be seen in operation at live steam festivals and traction engine rallies such as the Great Dorset Steam Fair in England, and the Western Minnesota Steam Threshers Reunion in northwest Minnesota.

Musical References

Irish songwriter John Duggan immortalised the threshing machine in a song *The Old Thrashing Mill*. The song has been recorded by Foster and Allen and Brendan Shine.

On the Alan Lomax collection Songs of Seduction (Rounder Select, 2000), there's a bawdy Irish folk song called "The Thrashing Machine" sung by tinker Annie O'Neil, as recorded in the early 20th Century.

In his film score for "Of Mice and Men" (1939) and consequently in his collection "Music for the Movies" (1942), American composer Aaron Copland titled a section of the score "Threshing Machines," to suit a scene in the Lewis Milestone film where Curley is threatening Slim over giving May a puppy, when many of the itinerant worker men are standing around or working on threshers.

In the song Thrasher from the album Rust Never Sleeps, Neil Young compares the modern threshing machine's technique of separating wheat from wheat stalks to the natural forces of time that separate close friends from one another.

Threshing machines appear in Twenty One Pilots' music video for the song House of Gold.

Hog Oiler

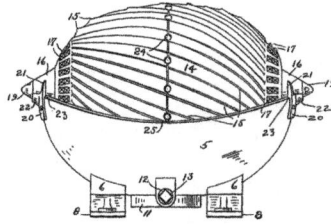

Hog oiler patent image U.S. Patent 1,241,023

A hog oiler was a mechanical device employed on farms to be used by hogs to provide relief from insects and offer skin protection. It consisted of a reservoir to hold oil, and a means to distribute the oil onto the hog, often via grooved wheels or cylinders. Hogs seeking relief would rub up against a wheel (or cylinder) causing it to rotate and dispense oil onto their bodies.

History

In the late 1800s and early 1900s hog lice and hog cholera could prove devastating to livestock producers and the hog oiler was seen as one way to combat the problem. The first known patent for a hog oiler device was issued in 1902 by the U.S. Patent Office, however the era of innovation for this device was mainly the years 1913-1923. According to *Goodbye Mr. Louse*, a book by Robert Rauhauser, there may have many as 157 different patented models of hog oilers, but collectors today estimate there could have been as many as 600 manufacturers, most going unpatented.

Prices for hog oilers would range anywhere from four dollars for a small model to twelve dollars and even higher for larger ones with more features. The same companies that manufactured the oilers would often sell special medicated oil to be used with the device, offering further protection. Many farmers however simply chose to use recycled oil or made their own cheaper versions. The U.S. government, while being willing to issue patents, suggested that the oilers might be less than effective, according to collector Bob Coates in *Farm Collector* magazine. "They (the government) recommended mopping or dipping (the hogs) instead" said Coats.

Companies throughout the midwestern U.S., such as Lisle Manufacturing of Clarinda, Iowa, offered farmers and hog producers a variety of styles and sizes including fence-mounted, stand-alone, walk-through and ratchet-governed. However the most common were double-wheel models known as Colubians and Sipes. Often made of cast iron, the early hog oilers could be quite heavy, with some models weighing as much as 150 pounds. Later models from the 1920s onward would mostly be constructed of cheaper, lighter steel and sheet metal. Other known manufacturers of hog oilers included the National Oiler Company of Richmond, Indiana, Rowe Manufacturing, Galesburg, Illinois, and Starbuck Manufacturing, Illinois Implement Company, and O.H.C.Manufacturing, all of Peoria, Illinois.

Collectibles

Antique "watermelon type" hog oiler.

World War II led to a small but growing collectors market for hog oilers today. The cast iron models are considered the most desirable antiques since many originals were gathered up in war scrap iron drives and destroyed. Hog oiler collectors come from all locations and ages, some as young as 11 years old. Generally the "hog belt" of the upper midwest provides the majority of hog oiler finds today—states such as Iowa, Illinois, Indiana, Nebraska and parts of Kansas and Missouri. Prices paid by collectors can vary widely, based on size, functions, and condition of the machine. Smaller hog oilers can be purchased for sometimes thirty dollars or less, while the rarest models can have asking prices in the many thousands of dollars. A small cast iron "wheel" type hog oiler that appeared nearly unused was featured on the History Channel program Pawn Stars on August 8, 2011. It sold for $100.

Baler

A Claas large round baler

An Abbriata Roto-Baler ejecting a net-wrapped small round bale

A baler, most often called a hay baler is a piece of farm machinery used to compress a cut and raked crop (such as hay, cotton, flax straw, salt marsh hay, or silage) into compact bales that are easy to handle, transport, and store. Often bales are configured to dry and preserve some intrinsic (e.g. the nutritional) value of the plants bundled. Several different types of balers are commonly used, each producing a different type of bale – rectangular or cylindrical, of various sizes, bound with twine, strapping, netting, or wire.

Industrial balers are also used in material recycling facilities, primarily for baling metal, plastic, or paper for transport.

History

Before the 19th century, hay was cut by hand and most typically stored in haystacks using hay forks to rake and gather the scythed grasses into optimal sized heaps — neither too large (promoting conditions that might create spontaneous combustion, nor too small, so much of the pile is susceptible to rotting. These haystacks lifted most of the plant fibers up off the ground, letting air in and water drain out, so the grasses could dry and cure, to retain nutrition for livestock feed at a later time. In the 1860s mechanical cutting devices were developed; from these came the modern devices including mechanical mowers and balers. In 1872 a reaper that used a knotter device to bundle and bind hay was invented by Charles Withington; this was commercialized in 1874 by Cyrus McCormick. In 1936, Innes invented an automatic baler that tied bales with twine using Appleby-type knotters from a John Deere grain binder; an improved version patented by Ed Nolt in 1939 was more reliable and became commonly used.

Round Baler

The most common type of baler in industrialized countries today is the round baler. It produces cylinder-shaped "round" or "rolled" bales. The design has a "thatched roof" effect that withstands weather well. Grass is rolled up inside the baler using rubberized belts, fixed rollers, or a combination of the two. When the bale reaches a predetermined size, either netting or twine is wrapped around it to hold its shape. The back of the baler swings open, and the bale is discharged. The bales are complete at this stage, but they may also be wrapped in plastic sheeting by a bale wrapper, either to keep hay dry when stored outside or convert damp grass into silage. Variable-chamber large round balers typically produce bales from 48 to 72 inches (120 to 180 cm) in diameter and up to 60 inches (150 cm) in width. The bales can weigh anywhere from 1,100 to 2,200 pounds (500 to 1,000 kg), depending upon size, material, and moisture content. Common modern small round balers (also called "mini round balers" or "roto-balers") produce bales 20 to 22 inches (51 to 56 cm) in diameter and 20.5 to 28 inches (52 to 71 cm) in width, generally weighing from 40 to 55 pounds (18 to 25 kg).

Allis Chalmers Rotobaler

Originally conceived by Ummo Luebben circa 1910, the first round baler did not see production until 1947 when Allis-Chalmers introduced the Roto-Baler. Marketed for the water-shedding and light weight properties of its hay bales, AC had sold nearly 70,000 units by the end of production in 1960. The next major innovation began in 1965 when a graduate student at Iowa State University, Virgil Haverdink, sought out Wesley F. Buchele, a professor of Agricultural Engineering, seeking a research topic for a master thesis. Over the next year Buchele and Haverdink developed a new design for a large round baler, completed and tested in 1966, and thereafter dubbed the Buchele-Haverdink large round baler. The large round bales were about 5 ft in diameter, 7 ft long, and they weighed about 600 lb after they dried—about 5 lb/ft^3. The design was promoted as a "Whale of a Bale" and Iowa State University now explains the innovative design as follows:

Round baler dumping a fresh bale

"Farmers were saved from the backbreaking chore of slinging hay bales in the 1960s when Iowa State agricultural engineering professor Wesley Buchele and a group of student researchers invented a baler that produced large, round bales that could be moved by tractor. The baler has become the predominant forage-handling machine in the United States."

In the summer of 1969, the Australian Econ Fodder Roller baler came out, a design that made a 300-lb ground-rolled bale. In September of that same year The Hawkbilt Company of Vinton, Iowa, contacted Dr. Buchele about his design, then fabricated a large ground-rolling round baler which baled hay that had been laid out in a windrow, and began manufacturing large round balers in 1970. In 1972, Gary Vermeer of Pella, Iowa, designed and fabricated a round baler after the design of the A-C Roto-Baler, and the Vermeer Company began selling its model 605 - the first modern round baler. The Vermeer design used belts to compact hay into a cylindrical shape as is seen today. In the early 1980s, collaboration between Walterscheid and Vermeer produced the first effective uses of CV joints in balers, and later in other farm machinery. Due to the heavy torque required for such equipment, double Cardan joints are primarily used. Former Walterscheid engineer Martin Brown is credited with "inventing" this use for universal joints.

By 1975, fifteen American and Canadian companies were manufacturing large round balers.

Transport, Handling, and Feeding

Short-haul Transport and on-field Handling

A large round bale

Due to the ability for round bales to roll away on a slope, they require specific treatment for safe transport and handling. Small round bales can typically be moved by hand or with lower-powered equipment. Large round bales, due to their size and weight (they can weigh a ton or more) require special transport and moving equipment.

The most important tool for large round bale handling is the bale spear or spike, which is usually mounted on the back of a tractor or the front of a skid-steer. It is inserted into the approximate center of the round bale, then lifted and the bale is hauled away. Once at the destination, the bale is set down, and the spear pulled out. Careful placement of the spear in the center is needed or the bale can spin around and touch the ground while in transport, causing a loss of control. When used for wrapped bales that are to be stored further, the spear makes a hole in the wrapping that must be sealed with plastic tape to maintain a hermetic seal.

Alternatively, a grapple fork may be used to lift and transport large round bales. The grapple fork is a hydraulically driven implement attached to the end of a tractor's bucket loader. When the hydraulic cylinder is extended, the fork clamps downward toward the bucket, much like a closing hand. To move a large round bale, the tractor approaches the bale from the side and places the bucket underneath the bale. The fork is then clamped down across the top of the bale, and the bucket is lifted with the bale in tow. Grab hooks installed on the bucket of a tractor are another tool used to handle round bales, and be done by a farmer with welding skills by welding two hooks and a heavy chain to the outside top of a tractor front loader bucket.

Long-haul Transport

The rounded surface of round bales poses a challenge for long-haul, flat-bed transport, as they could roll off of the flat surface if not properly supported. This is particularly the case with large round bales; their size makes them difficult to flip, so it may not be feasible to flip many of them onto the flat surface for transport and then re-position them on the round surface at the destination. One option that works with both large and small round bales is to equip the flat-bed trailer with guard-rails at either end, which prevent bales from rolling either forward or backward. Another solution is the saddle wagon, which has closely spaced rounded saddles or support posts in which round bales sit. The tall sides of each saddle prevent the bales from rolling around while on the wagon, as the bale settles down in between posts. On 3 September 2010, on the A381 in Halwell near Totnes, Devon, UK an early member of British rock group ELO Mike Edwards was killed when his van was crushed by a large round bale. The cellist, 62, died instantly when the 600 kilograms (1,300 lb) bale fell from a tractor on nearby farmland before rolling onto the road and crushing his van.

Feeding

A large round bale can be directly used for feeding animals by placing it in a feeding area, tipping it over, removing the bale wrap, and placing a protective ring (a *ring feeder*) around the outside so that animals don't walk on hay that has been peeled off the outer perimeter of the bale. The round baler's rotational forming and compaction process also enables both large and small round bales to be fed out by unrolling the bale, leaving a continuous flat strip in the field or behind a feeding barrier.

Silage or Haylage

A recent innovation in hay storage has been the development of the silage or haylage bale, which is a high-moisture bale wrapped in plastic film. These are baled much wetter than hay bales, and are usually smaller than hay bales because the greater moisture content makes them heavier and harder to handle. These bales begin to ferment almost immediately, and the metal bale spear stabbed into the core becomes very warm to the touch from the fermentation process.

Silage or haylage bales may be wrapped by placing them on a rotating bale spear mounted on the rear of a tractor. As the bale spins, a layer of plastic cling film is applied to the exterior of the bale. This roll of plastic is mounted in a sliding shuttle on a steel arm and can move parallel to the bale axis, so the operator does not need to hold up the heavy roll of plastic. The plastic layer extends over the ends of the bale to form a ring of plastic approximately 12 inches (30 cm) wide on the ends, with hay exposed in the center.

To stretch the cling-wrap plastic tightly over the bale, the tension is actively adjusted with a knob on the end of the roll, which squeezes the ends of the roll in the shuttle. In this example wrapping video, the operator is attempting to use high tension to get a flat, smooth seal on the right end. However, the tension increases too much and the plastic tears off. The operator recovers by quickly loosening the tension and allows the plastic to feed out halfway around the bale before reapplying the tension to the sheeting.

These bales are placed in a long continuous row, with each wrapped bale pressed firmly against all the other bales in the row before being set down onto the ground. The plastic wrap on the ends of each bale sticks together to seal out air and moisture, protecting the silage from the elements. The end-bales are hand-sealed with strips of cling plastic across the opening.

Large rectangular baler.

The airtight seal between each bale permits the row of round bales to ferment as if they were in a silo bag, but they are easier to handle than a silo bag, as they are more robust and compact. The plastic usage is relatively high, and there is no way to reuse the silage-contaminated plastic sheeting, although it can be recycled or used as a fuel source via incineration. The wrapping cost is approximately US$5 per bale.

An alternative form of wrapping uses the same type of bale placed on a bale wrapper, consisting of pair of rollers on a turntable mounted on the three-point linkage of a tractor. It is then spun about two axes while being wrapped in several layers of cling-wrap plastic film. This covers the ends and sides of the bale in one operation, thus sealing it separately from other bales. The bales are then moved or stacked using a special pincer attachment on the front loader of a tractor, which does not damage the film seal. They can also be moved using a standard bale spike, but this punctures the airtight seal, and the hole in the film must be repaired after each move.

Plastic-wrapped bales must be unwrapped before being fed to livestock to prevent accidental ingestion of the plastic. Like round hay bales, silage bales are usually fed using a *ring feeder*.

Large Rectangular Baler

Another type of baler in common use, in some areas, will produce large rectangular bales, each bound with a half dozen or so strings of twine, which are then knotted. Such bales are highly compacted and generally weigh somewhat more than round bales. The large rectangular bales are several times larger than the similar small bales. In the prairies of Canada, the large rectangular balers are also called "prairie raptors".

Rectangular Bale Handling and Transport

Large rectangular bales in a field, Charente, France. Sizes of stacks of baled hay need carefully managed, as the curing process is exothermic and the built up heat around internal bales can reach ignition temperatures in the right weather history and atmospheric conditions. Building a deep stack either too wide, or too high increases the risk of spontaneous ignition.

Rectangular bales are easier to transport than round bales, since there is little risk of the bale rolling off the back of a flatbed trailer. The rectangular shape also saves space and allows a complete solid slab of hay to be stacked for transport and storage. Most balers allow adjustment of length and it is common to produce bales of twice the width, allowing stacks with brick-like alternating groups overlapping the row below at right angles, creating a strong structure.

They are well-suited for large scale livestock feedlot operations, where many tons of feed are rationed every hour. Most often, they are baled small enough that one person can carry or toss them where needed.

Due to the huge rectangular shape, large spear forks, or squeeze grips are mounted to heavy lifting machinery, such as large fork lifts, tractors equipped with front end loaders, telehandlers, hay squeezes or wheel loaders, to lift these bales.

Small Rectangular Baler

A small square baler

A type of baler that produces small rectangular (often called "square") bales was once the most prevalent form of baler, but is less common today. It is primarily used on small acreages where large equipment is impractical, and also for the production of hay for small operations, particularly horse owners who may not have access to the specialized feeding machinery used for larger baled. Each bale is about 15 in x 18 in x 40 in (40 x 45 x 100 cm). The bales are usually wrapped with two, but sometimes three, or more strands of knotted twine. The bales are light enough for one person to handle, about 45 to 60 pounds (20 to 27 kg).

To form the bale, the material to be baled, (which is often hay or straw) in the windrow is lifted by tines in the baler's *reel*. This material is then *packed* into the bale chamber, which runs the length of one side of the baler (normally the right hand side when viewed from the front). A combination plunger and knife move back and forth in the front of this chamber, with the knife closing the door into the bale chamber as it moves backwards. The plunger and knife are attached to a heavy asymmetrical flywheel to provide extra force as they pack the bales. A measuring device—normally a spiked wheel that is turned by the emerging bales—measures the amount of material that is being compressed and, at the appropriate length it triggers the *knotters* that wrap the twine around the bale and tie it off. As the next bale is formed the tied one is driven out of the rear of the baling chamber onto the ground or onto a special wagon or collecting sled hooked to the rear of the baler. This process continues as long as there is material to be baled, and twine to tie it with. The bales emerge with four sides. The twine runs, in two parallel loops, around the wider sides. Of the two narrower sides, there is a cut side and a dull side, and when stacked for storage or transport the bales are normally positioned with the cut side facing outwards.

This form of bale is not much used in large-scale commercial agriculture, because of the costs involved in handling many small bales. However, it enjoys some popularity in small-scale, low-mechanization agriculture and horse-keeping. Besides using simpler machinery and being easy to handle, these small bales can also be used for insulation and building materials in straw-bale construction. Square bales may generally weather better than round bales because a more much dense stack can be put up. However, they don't shed water as round bales do. Convenience is also a major factor in farmers deciding to continue putting up square bales, as they make feeding and bedding in confined areas (stables, barns, etc.) much easier.

Many of these older balers are still to be found on farms today, particularly in dry areas, where bales can be left outside for long periods.

The automatic-baler for small square bales took on most of its present form in 1940. It was first manufactured by the *New Holland Ag* and it used a small petrol engine to provide operating power. It is based on a 1937 invention for a twine-tie baler with automatic pickup.

Wire Balers

Stationary baler

Bales prior to 1937 were manually wire-tied with two baling wires. Even earlier, the baler was a stationary implement, driven with a tractor or stationary engine using a belt on a belt pulley, with the hay being brought to the baler and fed in by hand. Later, balers were made mobile, with a 'pickup' to gather up the hay and feed it into the chamber. These often used air cooled gasoline engines mounted on the baler for power. The biggest change to this type of baler since 1940 is being powered by the tractor through its power take-off (PTO), instead of by a built-in internal combustion engine.

In present-day production, small square balers can be ordered with twine knotters or wire tie knotters.

Not all stationary wire tying balers used 2 wires. It was not uncommon for the larger bale size (usually 17" x 22") machines to use 'boards' that had three slots for wires and hence tied three wires per bale. Most North American manufacturers produced these machines as either regular models or as size options. 'Small square' three wire tying pick-up balers were available from the early 1930s, principally from J. I. Case & Co. and Ann Arbor. These machines were hand tying and hand threading machines. Although New Holland credits itself with inventing the 'successful small square twine tying baler', it produced such machines for the first time in 1940 after acquiring Ed Nolte and his baler. This baler baled successfully from 1937 onwards. Certainly the quality of the New Holland machines, popularised twine tying hay balers. In Europe, in as early as 1939, both Claas of Germany and Rousseau SA of France had automatic twine tying pick-up balers. Most of these produced low density bales though. The first successful pick-up balers were made by the Ann Arbor Company in 1929. Ann Arbor were acquired by the Oliver Farm Equipment Company in 1943. Despite their head start on the rest of the field, no Ann Arbor balers carried automatic knotters or twisters. Oliver introduced these in 1949.

Square/Wire Bale History

Pickup and Handling Methods

In the 1940s most farmers would bale hay in the field with a small tractor with 20 or less horsepower, and the tied bales would be dropped onto the ground as the baler moved through the field. Another team of workers with horses and a flatbed wagon would come by and use a sharp metal hook to grab the bale and throw it up onto the wagon while an assistant stacks the bale, for transport to the barn.

A later time-saving innovation was to tow the flatbed wagon directly behind the baler, and the bale would be pushed up a ramp to a waiting attendant on the wagon. The attendant hooks the bale off the ramp and stacks it on the wagon, while waiting for the next bale to be produced.

Eventually, as tractor horsepower increased, the thrower-baler became possible, which eliminated the need for someone to stand on the wagon and pick up the finished bales. The first thrower mechanism used two fast-moving friction belts to grab finished bales and throw them at an angle up in the air onto the bale wagon. The bale wagon was modified from a flatbed into a three-sided skeleton frame open at the front, to act as a catcher's net for the thrown bales.

As tractor horsepower further increased, the next innovation of the thrower-baler was the hydraulic tossing baler. This employs a flat pan behind the bale knotter. As bales advance out the back of the baler, they are pushed onto the pan one at a time. When the bale has moved fully onto the pan, the pan suddenly pops up, pushed by a large hydraulic cylinder, and tosses the bale up into the wagon like a catapult.

The pan-thrower method puts much less stress on the bales compared to the belt-thrower. The friction belts of the belt-thrower stress the twine and knots as they grip the bale, and would occasionally cause bales to break apart in the thrower or when the bales landed in the wagon.

Square bale stacker

Bales may be picked up from the field and stacked using a self-powered machine called a *bale stacker*, *bale wagon* or *harobed*. There are several designs and sizes. One type picks up square bales are dropped by the baler with the strings facing upward. The stacker will drive up to each bale, pick it up and set it on a three-bale-wide table (the strings are now facing upwards). Once three bales are on the table, the table lifts up and back, causing the three bales to face strings to the side again; this happens three more times until there are 16 bales on the main table. This table will lift like the smaller one, and the bales will be up against a vertical table. The machine will hold 160 bales (ten tiers); usually there will be cross-tiers near the center to keep the stack from swaying or collapsing if any weight is applied to the top of the stack. The full load will be transported to a barn; the whole rear of the stacker will tilt upwards until it is vertical. There will be two pushers that will extend through the machine and hold the bottom of the stack from being pulled out from the stacker while it is driven out of the barn.

A smaller type of stacker

In Britain (if small square bales are still to be used), they are usually collected as they fall out of the baler in a *bale sledge* dragged behind the baler. This has four channels, controlled by automatic mechanical balances, catches and springs, which sort each bale into its place in a square *eight*. When the sledge is full, a catch is tripped automatically, and a door at the rear opens to leave the eight lying neatly together on the ground. These may be picked up individually and loaded by hand, or they may be picked up all eight together by a *bale grab* on a tractor, a special front loader consisting of many hydraulically powered downward-pointing curved spikes. The square eight will then be stacked, either on a trailer for transport, or in a roughly cubic field stack eight or ten layers high. This cube may then be transported by a large machine attached to the three-point hitch behind a tractor, which clamps the sides of the cube and lifts it bodily.

Storage Methods

Before electrification occurred in rural parts of the United States in the 1940s, some small dairy farms would have tractors but not electric power. Often just one neighbor who could afford a tractor would do all the baling for surrounding farmers still using horses.

To get the bales up into the hayloft, a pulley system ran on a track along the peak of the barn's hayloft. This track also stuck a few feet out the end of the loft, with a large access door under the track. On the bottom of the pulley system was a bale spear, which is pointed on the end and has retractable retention spikes.

A flatbed wagon would pull up next to the barn underneath the end of the track, the spear lowered down to the wagon, and speared into a single bale. The pulley rope would be used to manually lift the bale up until it could enter the mow through the door, then moved along the track into the barn and finally released for manual stacking in tight rows across the floor of the loft. As the stack filled the loft, the bales would be lifted higher and higher with the pulleys until the hay was stacked all the way up to the peak.

When electricity arrived, the bale spear, pulley and track system were replaced by long motorized bale conveyors known as hay elevators. A typical elevator is an open skeletal frame, with a chain that has dull 3-inch (76 mm) spikes every few feet along the chain to grab bales and drag them along. One elevator replaced the spear track and ran the entire length of the peak of the barn. A second elevator was either installed at a 30-degree slope on the side of the barn to lift bales up to the peak elevator, or used dual front-back chains surrounding the bale to lift bales straight up the side of the barn to the peak elevator.

A bale wagon pulled up next to the lifting elevator, and a farm worker placed bales one at a time onto the angled track. Once bales arrived at the peak elevator, adjustable tipping gates along the length of the peak elevator were opened by pulling a cable from the floor of the hayloft, so that bales tipped off the elevator and dropped down to the floor in different areas of the loft. This permitted a single elevator to transport hay to one part of a loft and straw to another part.

This complete hay elevator lifting, transport, and dropping system reduced bale storage labor to a single person, who simply pulls up with a wagon, turns on the elevators and starts placing bales on it, occasionally checking to make sure that bales are falling in the right locations in the loft.

The neat stacking of bales in the loft is often sacrificed for the speed of just letting them fall and roll down the growing pile in the loft, and changing the elevator gates to fill in open areas around the loose pile. But if desired, the loose bale pile dropped by the elevator could be rearranged into orderly rows between wagon loads.

Usage Once in the Barn

The process of retrieving bales from a hayloft has stayed relatively unchanged from the beginning of baling. Typically workers were sent up into the loft, to climb up onto the bale stack, pull bales off the stack, and throw or roll them down the stack to the open floor of the loft. Once the bale is down on the floor, workers climb down the stack, open a cover over a bale chute in the floor of the loft, and push the bales down the chute to the livestock area of the barn.

Most barns were equipped with several chutes along the sides and in the center of the loft floor. This permitted bales to be dropped into the area where they were to be used. Hay bales would be dropped through side chutes, to be broken up and fed to the cattle. Straw bales would be dropped down the center chute, to be distributed as bedding in the livestock standing/resting areas.

Traditionally multiple bales were dropped down to the livestock floor and the twine removed by hand. After drying and being stored under tons of pressure in the haystack, most bales are tightly compacted and need to be torn apart and fluffed up for use.

One recent method of speeding up all this manual bale handling is the bale shredder, which is a large vertical drum with rotary cutting/ripping teeth at the base of the drum. The shredder is placed under the chute and several bales dropped in. A worker then pushes the shredder along the barn aisle as it rips up a bale and spews it out in a continuous fluffy stream of material.

Field of hay bales, lines in field made by baler

Industrial Balers

A specialized baler designed to compact stretch wrap.

Industrial balers are typically used to compact similar types of waste, such as office paper, Corrugated fiberboard, plastic, foil and cans, for sale to recycling companies. These balers are made of steel with a hydraulic ram to compress the material loaded. Some balers are simple and labor-intensive, but are suitable for smaller volumes. Other balers are very complex and automated, and are used where large quantities of waste are handled.

Bulk Tank

In dairy farming a bulk milk cooling tank is a large storage tank for cooling and holding milk at a cold temperature until it can be picked up by a milk hauler. The bulk milk cooling tank is an important piece of dairy farm equipment. It is usually made of stainless steel and used every day to store the raw milk on the farm in good condition. It must be cleaned after each milk collection. The milk cooling tank can be the property of the farmer or be rented from a dairy plant.

Bulk Tank Types

Different types of milk cooling tanks

Raw milk producers have a choice of either open (from 150 to 3000 litres) or closed (from 1000 to 10000 litres) tanks. The cost can vary considerably, depending on manufacturing norms and whether a new or second hand tank is purchased.

Milk silos (10,000 litres and plus) are suitable for the very large producer. These are designed to be installed outside and adjacent to the dairy, all controls and the milk outlet pipe being situated in the dairy.

Tank Construction

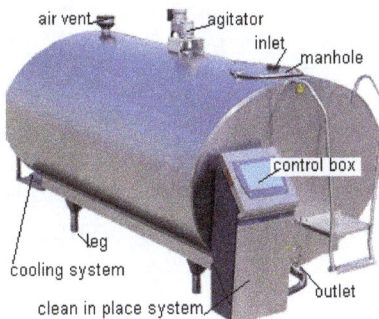

Bulk milk cooling tank description

A milk cooling tank, also known as a bulk tank or milk cooler, consists of an inner and an outer tank, both made of high quality stainless steel.

The space between the outer tank and the inner tank is isolated with polyurethane foam. In case of a power failure with an outside temperature of 30°C, the content of the tank will warm up only 1°C in 24 hours.

To facilitate an adequate and rapid cooling of the entire content of a tank, every tank is equipped with at least one agitator. Stirring the milk ensures that all milk inside the tank is of the same temperature and that the milk stays homogeneous.

On top of every closed milk cooling tank is a manhole of about 40 centimetres diameter. This enables thorough cleaning and inspection of the inner tank if necessary. The manhole is covered by a lid and sealed watertight with a rubber ring. Also on top are 2 or 3 small inlets. One is covered with an air-vent, the other(s) can be used to pump milk into the tank.

A milk cooling tank usually stands on 4, 6, or 8 adjustable legs. The built-in tilt of the inner tank ensures that even the last drop of milk will eventually flow to the outlet.

At the bottom, every milk cooling tank has a threaded outlet, usually including a valve.

All tanks have a thermometer, allowing for immediate inspection of the inner temperature.

Most tanks include an automatic cleaning system. Using hot and cold water, an acid and/or alkaline cleaning fluid, a pump and a spray lance will clean the inner tank, ensuring an hygienic inner environment each time the tank is emptied.

Almost every tank has a control box. It manages the cooling process by use of a thermostat. The user can turn the system on and off, allow for extra and immediate stirring, start the cleaning routine, and reset the entire system in case of a failure.

New and bigger milk cooling tanks are now being equipped with monitoring and alarm systems. These systems guard temperature of the milk inside the tank, check the functioning of the agitator, the cooling unit and temperature of the cleaning water. In case of malfunctioning of any of these functions, the alarm will activate. The monitoring system will also keep a record of the temperature and of all malfunctions for a given period.

Bulk Tank Manufacturing Norms

Norms define among other criteria: insulation, milk agitation, cooling power required, variations in milk quantity measurement, calibration, ... Some are more demanding than others.

- ISO standard 5708 (*Refrigerated bulk milk tanks*), published in 1983

- European standard EN 13732 (*Food processing machinery – Bulk milk coolers on farms – Requirements for construction, performance, suitability for use, safety and hygiene*), published in 2003, updated in 2009

- Northern American sanitary standard 3A 13-11 (*3-A Sanitary Standards for Farm Milk Cooling and Holding Tanks, 13-11*), effective July 23, 2012 www.3-a.org

Bulk Tank Outlet Standards

Swedish outlet (SMS 1145), German outlet (DIN 11851), English RJT (BS 4825), IDF (ISO 2853), tri-clamp (ISO 2852), Danish outlet (DS 722), ... can be found, not to mention different diameters. They vary from country to country. Non standard outlets make the milk collection process difficult, as the operator needs to adapt to each different standard/diameter.

Cooling Systems

There are two primary methods of cooling milk entering the bulk tank, each with its own advantages and disadvantages. The tank capacity and type will depend on herd size, calving pattern, frequency of milk collection, required milk quality, energy and water availability and future plans for development.

Direct Expansion

A bulk tank with direct expansion cooling has pipes or pillow plates carrying refrigerant which are welded directly to the exterior of the milk chamber. A layer of insulation covers the exterior of the milk tank and the cooling lines, with an exterior metal shell over the insulation.

Direct expansion cooling cannot run when the tank is empty or the inside walls of the tank would freeze. Instead, the tank is rapidly cooled as warm milk first enters the tank, and then the tank is cooled slowly just to maintain a low storage temperature. The rapid cooling during milking requires very large refrigeration compressors and condenser radiators to quickly expel heat from the milk, and is better suited for very large farming operations where three-phase electric power is available to operate the high-power cooling system.

Ice Bank

A bulk tank using an Ice Builder or Ice Bank immerses the bottom of the inner milk chamber in an open pool of water with copper tubes containing refrigerant suspended in the water. Between milkings, a small low-power cooling system slowly builds up a coating of ice around the copper tubes, and prevents icing of the pool over by continuously circulating the water in the pool. After the ice has achieved a thickness of 2-3 inches, the cooling system stops running.

During milking, the milk entering the tank is primarily cooled by circulating the water in the pool around the walls of the inner milk chamber, and the melting of the ice. After the ice has melted sufficiently the cooling system restarts to assist the ice bank and restart the ice building.

Ice bank bulk tanks are better suited for small family farm operations where only single-phase electric power is available, and high-power cooling systems would be either too expensive or difficult to install.

Milk Pre-cooling

For energy savings and quality reasons it is advisable to pre-cool the milk before it enters the tank using a plate or a tube cooler (shell and tube heat exchanger) supplied with chilled water from the well water, the ice builder or the condensing unit. The quicker milk is cooled after leaving the cow the better. This system achieves most of the cooling before the milk enters the tank, so that chilled milk, rather than warm milk, is being added to the already cooled milk in the tank.

Cooling Temperature

Generic temperature for milk storage is 3 to 4°C. For raw milk cheese manufacturing, it would be advisable to keep the milk at 12°C, as milk characteristics will be kept in a better state.

The milk cooling tank is usually not completely filled at once. A 2 milking tank is designed to cool 50% of its capacity at once. A 4 milking tank is designed to cool 25% of its capacity at once, and a 6 milking tank is designed to cool 16.7% of its capacity at once.

The cooling performance depends on the number of milking it takes to completely fill the tank, the ambient temperature and the cooling time.

Bulk Tank Cleaning Systems

There are two primary methods of cleaning bulk tanks, via manual scrubbing or automatic washing. Both methods generally use four steps to clean the tank:

- prerinsing with water to wet the surface and rinse off remaining milk residue

- washing with hot soapy water

- rinsing with water to remove the soap

- final *sanitizing* rinse with an approved bulk tank sanitizer solution

Manual Scrubbing

Manual scrubbing requires the bulk tank to have large hinged covers that can be lifted open to permit easy access to the interior surfaces of the tank. It tends to be much more thorough than automatic methods since it permits the tank to be carefully inspected during the washing process. If the tank is not found to be cleaned well enough, a troublesome area can be given additional cleansing attention.

Manual Scrubbing Limitations

This job is difficult to perform for very large tanks, and becomes more difficult as the overall cross-section or diameter of the tank increases, requiring either a longer brush or a raised work platform around the tank to lift the cleaning worker to reach over the side of a tall tank.

Automatic Washing

Automatic bulk tank washing and are normally activated by the milk collection truck driver after each milk collection. The cleaning system operates similar to a consumer dishwasher and consists of one or more free-spinning high-pressure spray nozzles with tangential jets, with the spray nozzle mounted on the end of a flexible whip suspended down into the center of the interior. As the cleaning solution sprays out of the jet, the force of the expelled water causes the jet to spin around and the whip to wildly swing back and forth, spraying the cleaning solution randomly all over the interior of the tank.

Automatic Washing Limitations

Because no physical scrubbing occurs with automatic wash systems, the cleanser relies on surfactants and detergents to dissolve the fats left on the interior of the tank by the cream in the milk. However, this is not sufficient to remove milkstone buildup, and the tank may need to be washed occasionally with milkstone remover to remove this scale buildup that can harbor bacteria and contaminants.

Automatic scrubbing only cleans the interior of the tank. It is not capable of cleaning the exterior of the tank, and it does not do a good job of washing around the cover seals. While it is possible to just clean the interior and call it good enough, it does not provide the maximum sanitation of manually washing down the exterior of the tank following or during the automatic wash process. Also, some components that contact the milk such as the drain valve cannot be properly cleaned automatically without disassembling the valve and retaining washer and directly scrubbing in soapy water.

Operating Costs

Substantial reductions in running costs can be made when an ice builder is used in conjunction with off-peak electricity. Pre-cooling milk using a plate or a tube cooler supplied with mains or well water can also reduce costs and add to the cooling capacity of the tank.

Bulk tank condenser units, which are not an integral part of the tank, should be fitted in an adjacent, suitable and well ventilated place.

If at all possible, condenser units should not be fitted on a wall facing the sun. They should be installed in a way which allows them to draw in and discharge adequate quantities of air for efficient operation.

Bulk tank should be easily accessible by large bulk collection tankers and positioned so that the tanker approaches can be kept clean and free from cow traffic at all times.

Although tanks have been calibrated when first installed, bulk tank miscalibration is not uncommon and in some cases it can result in significant loss of income. Milk tanks calibrated on the low side, can cheat raw milk producers by up to 22 litres on each shipment. It is therefore advisable to re-calibrate a bulk tank.

Other Usage of Bulk Tanks

Stainless steel bulk tanks are also used to heat or cool a fluid or simply to keep it isolated and warm/cold. Because of the hygienical finishing of the inner and outer side of the tanks, almost any fluid can be stored: water, fruit juices, honey, wine, beer, ink, paint, cosmetics, aromatic food-additives, bacterial cultures, cleansers, oil, or blood.

References

- Wells, David A. (1891). Recent Economic Changes and Their Effect on Production and Distribution of Wealth and Well-Being of Society. New York: D. Appleton and Co. ISBN 0-543-72474-3.

- R. Goyal, Megh (2012). Management of drip/trickle or micro irrigation. Oakville, CA: Apple Academic Press. p. 104. ISBN 978-1926895123.

- Clark, Gregory (2007). A Farewell to Alms: A Brief Economic History of the World. Princeton University Press. p. 286. ISBN 978-0-691-12135-2.

- Fox Business News: "Machinery takes the place of migrants as Maine's blueberry harvest booms" September 06, 2015

- "No Hands Touch the Land: Automating California Farms" (PDF). California Agrarian Action Project: 20–28. July 1977. Retrieved 2015-04-25.

- U. S. Produce Industry and Labor: Facing the Future in a Global Industry By Linda Calvin retrieved September 28, 2013

- "Mulchers (Also referred to as Masticators or Brushcutters)". Forest Operations Equipment Catalog. Forests and Rangelands. Retrieved 7 July 2013.

- A.E. Araiza Arizona Daily Star (2010-05-06). "Mechanical grinder quickly creates firebreaks in Oracle". Azstarnet.com. Retrieved 2013-06-28.

- "Wildlife in "Edge Areas" Thrive When Habitats are Kept in Check". Forconstructionpros.com. 2011-01-04. Retrieved 2013-06-28.

- "Tamarac National Wildlife Refuge, Becker County, Minnesota | Woodcock population and young forest habitat management". Timberdoodle.org. Retrieved 2013-06-28.

Agricultural Robot: An Integrated Study

Agricultural robots are used in agriculture; they mainly serve in the harvesting stage of the process. These agricultural robots are designed specifically for replacing human labor. The kinds of robots used are fruit picking robots, driverless sprayers and sheep shearing robots. This chapter will provide an integrated understanding on agricultural robots.

Agricultural Robot

Agricultural robots or agbot is a robot deployed for agricultural purposes. The main area of application of robots in agriculture today is at the harvesting stage. A possible emerging application of robots or drones is for weed control.

General

Fruit picking robots, driverless tractor / sprayer, and sheep shearing robots are designed to replace human labor. In most cases, a lot of factors have to be considered (e.g., the size and color of the fruit to be picked) before the commencement of a task. Robots can be used for other horticultural tasks such as pruning, weeding, spraying and monitoring. Robots can also be used in livestock applications (livestock robotics) such as automatic milking, washing and castrating. Robots like these have many benefits for the agricultural industry, including a higher quality of fresh produce, lower production costs, and a smaller need for manual labor. They can also be used to automate manual tasks, such as weed or bracken spraying, where the use of tractors and other manned vehicles is too dangerous for the operators.

Designs

The mechanical design consists of an end effector, manipulator, and gripper. Several factors must be considered in the design of the manipulator, including the task, economic efficiency, and required motions. The end effector influences the market value of the fruit and the gripper's design is based on the crop that is being harvested.

End Effectors

An end effector in an agricultural robot is the device found at the end of the robotic arm, used for various agricultural operations. Several different kinds of end effectors have been developed. In an agricultural operation involving grapes in Japan, end effectors are used for harvesting, berry-thinning, spraying, and bagging. Each was designed according to the nature of the task and the shape and size of the target fruit. For instance, the end effectors used for harvesting were designed to grasp, cut, and push the bunches of grapes.

Berry thinning is another operation performed on the grapes, and is used to enhance the market value of the grapes, increase the grapes' size, and facilitate the bunching process. For berry thinning, an end effector consists of an upper, middle, and lower part. The upper part has two plates and a rubber that can open and close. The two plates compress the grapes to cut off the rachis branches and extract the bunch of grapes. The middle part contains a plate of needles, a compression spring, and another plate which has holes spread across its surface. When the two plates compress, the needles punch holes through the grapes. Next, the lower part has a cutting device which can cut the bunch to standardize its length.

For spraying, the end effector consists of a spray nozzle that is attached to a manipulator. In practice, producers want to ensure that the chemical liquid is evenly distributed across the bunch. Thus, the design allows for an even distribution of the chemical by making the nozzle to move at a constant speed while keeping distance from the target.

The final step in grape production is the bagging process. The bagging end effector is designed with a bag feeder and two mechanical fingers. In the bagging process, the bag feeder is composed of slits which continuously supply bags to the fingers in an up and down motion. While the bag is being fed to the fingers, two leaf springs that are located on the upper end of the bag hold the bag open. The bags are produced to contain the grapes in bunches. Once the bagging process is complete, the fingers open and release the bag. This shuts the leaf springs, which seals the bag and prevents it from opening again.

Gripper

The gripper is a grasping device that is used for harvesting the target crop. Design of the gripper is based on simplicity, low cost, and effectiveness. Thus, the design usually consists of two mechanical fingers that are able to move in synchrony when performing their task. Specifics of the design depend on the task that is being performed. For example, in a procedure that required plants to be cut for harvesting, the gripper was equipped with a sharp blade.

Manipulator

The manipulator allows the gripper and end effector to navigate through their environment. The manipulator consists of four-bar parallel links that maintain the gripper's position and height. The manipulator also can utilize one, two, or three pneumatic actuators. Pneumatic actuators are motors which produce linear and rotary motion by converting compressed air into energy. The pneumatic actuator is the most effective actuator for agricultural robots because of its high power-weight ratio. The most cost efficient design for the manipulator is the single actuator configuration, yet this is the least flexible option.

Applications

Robots have many fields of application in agriculture. Some examples and prototypes of robots include the Merlin Robot Milker, Rosphere, Harvest Automation, Orange Harvester, lettuce bot, and weeder. One case of a large scale use of robots in farming is the milk bot. It is widespread among British dairy farms because of its efficiency and nonrequirement to move. According to David Gardner (chief executive of the Royal Agricultural Society of England), a robot can complete a complicated task if its repetitive and the robot is allowed to sit in a single place. Furthermore, robots that work on repetitive tasks (e.g. milking) fulfill their role to a consistent and particular standard.

Another field of application is horticulture. One horticultural application is the development of RV 100 by Harvest Automation Inc. RV 100 is designed to transport potted plants in a greenhouse or outdoor setting. The functions of RV100 in handling and organizing potted plants include spacing capabilities, collection, and consolidation. The benefits of using RV100 for this task include high placement accuracy, autonomous outdoor and indoor function, and reduced production costs.

Fruit Picking

Orchard ladders in old farmstead apple orchard British Columbia, Canada, 2005

Fruit picking or fruit harvesting is a seasonal activity (paid or recreational) that occurs during harvest time in areas with fruit growing wild or being farmed in orchards.

Types of Fruit

Apple Picking

Apple picking in Styria

Apple picking is an activity found at apple farms. Apple orchards may be opened to the public, allowing consumers to pick their own apples or purchase pre-picked apples.

Although this is ultimately a method of purchasing apples, it is often a social activity as well. Apple picking is often a very popular dating ritual in the American Midwest. Apple orchards catering to a family outing will provide additional activities beyond the picking of apples. Many have petting zoos, restaurants and country shops that sell related products such as home-made jams and jellies. This aspect of the activity is especially popular in the Northeastern United States & Southern Ontario and Southern Québec in Canada.

The apples that fall off the trees are often used to make apple cider. Apple cider is a juice made grinding the apples, then pressing out the juice.

Workers

Fruit-growing polytunnels with caravans for the mainly Eastern European fruit workers conveniently parked behind, Perth and Kinross, Scotland, March 2009

Most of the fruit picking is done by migrant workers. Migrant workers are frequently used as they can be paid relatively low wages. In California, Mexican migrants are most frequently doing the work. There has been much controversy about replacing workers with automation. It puts many out of work.

In Australia and New Zealand a lot of fruit picking work is done by backpackers on a Working Holiday Visa. The Australian government encourages people on this visa to do this sort of work for a minimum of 3 months so they can add another year to their visa. This benefit is not for all parts of Australia, you must undertake work in selected post codes to be eligible for the extra year.

Automation

As labor costs are still quite expensive in fruit picking, robots are being designed that can replace humans for this kind of work. The research is still in full progress, especially as the robots need to be carefully designed so that they do not bruise the fruit while picking. One solution is the use of suction grippers, used on automated fruit picking machines manufactured, for example, by ACRO. Citrus fruit robot pickers have thus far been the focus of research and development, but cherry pickers are also being researched. Vision Robotics, in particular, has made several robots that are already capable of taking over the work.

Fruit Picking in Art

Lucas Cranach the Elder, Paradise (detail), 1530

Christian Berentz, Flowers, Fruit with a Woman Picking Grapes, 1696

Automatic Milking

Automatic milking is the milking of dairy animals, especially of dairy cattle, without human labour. Automatic milking systems (AMS), also called voluntary milking systems

(VMS), were developed in the late 20th century. They have been commercially available since the early 1990s. The core of such systems that allows complete automation of the milking process is a type of agricultural robot. Automated milking is therefore also called robotic milking. Common systems rely on the use of computers and special herd management software.

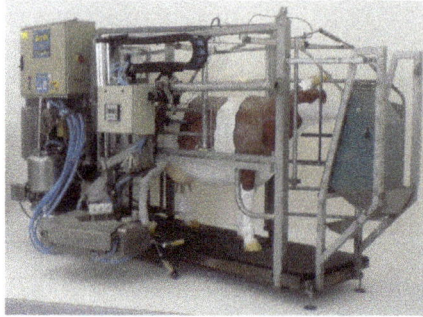

A Fullwood *Merlin* AMS unit from the 1990s, exhibit at the Deutsches Museum in Germany

Automation in Milking

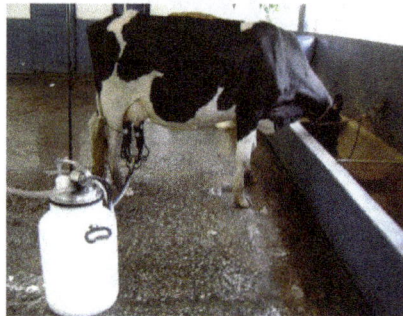

A cow and a milking machine – partial automation compared to hand milking

A rotary milking parlor – higher efficiency compared to stationary milking parlors, but still requiring manual labour with milking machines etc.

Basics – milking Process and Milking Schedules

The milking process is the collection of tasks specifically devoted to extracting milk from an animal (rather than the broader field of dairy animal husbandry). This process may be broken down into several sub-tasks: collecting animals before milking, routing animals into the parlour, inspection and cleaning of teats, attachment of milking equipment to teats, and often massaging the back of the udder to relieve any held back milk, extraction of milk, removal of milking equipment, routing of animals out of the parlour.

Maintaining milk yield during the lactation period (approximately 300 days) requires consistent milking intervals, usually twice daily and with maximum time spacing between milkings. In fact all activities must be scheduled around the milking process on the dairy farm. Such a milking routine imposes restrictions on time management and personal life of an individual farmer, as the farmer is committed to milking in the early morning and in the evening for seven days a week regardless of personal health, family responsibilities or social schedule. This time restriction is exacerbated for lone farmers and farm families if extra labour cannot easily or economically be obtained, and is a factor in the decline in small-scale dairy farming. Techniques such as once-a-day milking and voluntary milking have been investigated to reduce these time constraints.

Automation Progress in the 20th Century

To alleviate the labour involved in milking, much of the milking process has been automated during the 20th century: many farmers use semi-automatic or automatic cow traffic control (powered gates, etc.), the milking machine (a basic form was developed in the late 19th century) has entirely automated milk extraction, and automatic cluster removal is available to remove milking equipment after milking. Automatic teat spraying systems are available, however there is some debate over the cleaning effectiveness of these.

The final manual labour tasks remaining in the milking process were cleaning and inspection of teats and attachment of milking equipment (milking cups) to teats. Automatic cleaning and attachment of milking cups is a complex task, requiring accurate detection of teat position and a dextrous mechanical manipulator. These tasks have been automated successfully in the voluntary milking system (VMS), or automatic milking system (AMS).

Automatic Milking Systems (AMS)

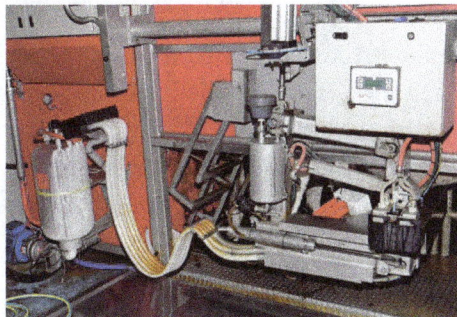

An older Lely Astronaut AMS unit at work (milking)

Since the 1970s, much research effort has been expended in investigating methods to alleviate time management constraints in conventional dairy farming, culminating in the development of the automated voluntary milking system. There is a video of the historical development of the milking robot at Silsoe Research Institute.

Voluntary milking allows the cow to decide her own milking time and interval, rather than being milked as part of a group at set milking times. AMS requires complete automation of the milking process as the cow may elect to be milked at any time during a 24-hour period.

The milking unit comprises a milking machine, a teat position sensor (usually a laser), a robotic arm for automatic teat-cup application and removal, and a gate system for controlling cow traffic. The cows may be permanently housed in a barn, and spend most of their time resting or feeding in the free-stall area. If cows are to be grazed as well, a selection gate is required to allow only those cows that have been milked to the outside pastures.

When the cow elects to enter the milking unit (due to highly palatable feed that she finds in the milking box), a cow ID sensor reads an identification tag (transponder) on the cow and passes the cow ID to the control system. If the cow has been milked too recently, the automatic gate system sends the cow out of the unit. If the cow may be milked, automatic teat cleaning, milking cup application, milking, and teatspraying takes place. As an incentive to attend the milking unit, concentrated feedstuffs needs to be fed to the cow in the milking unit.

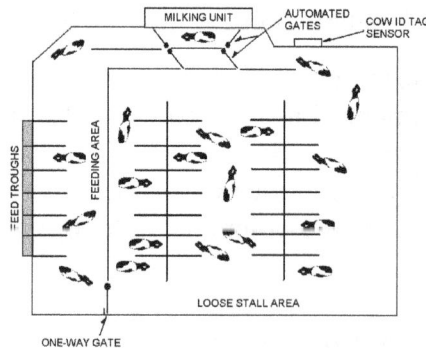

Typical VMS stall layout (forced cow traffic layout)

The barn may be arranged such that access to the main feeding area can only be obtained by passing the milking unit. This layout is referred to as *forced cow traffic*. Alternatively, the barn may be set up such that the cow always has access to feed, water, and a comfortable place to lie down, and is only motivated to visit the milking system by the palatable feed available there. This is referred to as *free cow traffic*.

The innovative core of the AMS system is the robotic manipulator in the milking unit. This robotic arm automates the tasks of teat cleaning and milking attachment and removes the final elements of manual labour from the milking process. Careful design of the robot arm and associated sensors and controls allows robust unsupervised performance, such that the farmer is only required to attend the cows for condition inspection and when a cow has not attended for milking.

Typical capacity for an AMS is 50–70 cows per milking unit. AMS usually achieve milking frequencies between 2 and 3 times per day, so a single milking unit handling 60 cows and milking each cow 3 times per day has a capacity of 7.5 cows per hour. This low capacity is convenient for lower-cost design of the robot arm and associated control system, as a window of several minutes is available for each cow and high-speed operation is not required.

AMS units have been available commercially since the early 1990s, and have proved relatively successful in implementing the voluntary milking method. Many of the research and developments have taken place in the Netherlands. The most farms with AMS are located in the Netherlands, and Denmark.

A new variation on the theme of robotic milking includes a similar robotic arm system, but coupled with a rotary platform, improving the number of cows that can be handled per robot arm.

Advantages

An AMS unit at work (teat cleaning)

- Elimination of labour - The farmer is freed from the milking process and associated rigid schedule, and labour is devoted to supervision of animals, feeding, etc.

- Milking consistency – The milking process is consistent for every cow and every visit, and is not influenced by different persons milking the cows. The four separate milking cups are removed individually, meaning that an empty quarter does not stay attached while the other three are finishing, resulting in less threat of injury. The newest models of automatic milkers can vary the pulsation rate and vacuum level based on milk flow from each quarter.

- Increased milking frequency – Milking frequency may increase to three times per day, however typically 2.5 times per day is achieved. This may result in less stress on the udder and increased comfort for the cow, as on average less milk is stored. Higher frequency milking increases milk yield per cow, however much of this increase is water rather than solids.

- Perceived lower stress environment – There is a perception that elective milking schedules reduce cow stress. A study found no decrease in stress between automatic and conventional milking.

- Herd management – The use of computer control allows greater scope for data collection. Such data allows the farmer to improve management through analysis of trends in the herd, for example response of milk production to changes in feedstuffs. Individual cow histories may also be examined, and alerts set to warn the farmer of unusual changes indicating illness or injury. Information gathering provides added value for AMS, however correct interpretation and use of such information is highly dependent on the skills of the user or the accuracy of computer algorithms to create attention reports.

Considerations and Disadvantages

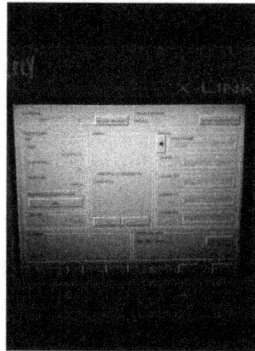

Touchscreen display of a milking robot

- Higher initial cost – AMS systems cost approximately €120,000 ($190,524) per milking unit as of 2003 (presuming barn space is already available for loose-stall housing). Equipment costs decreased from $175,000 for the first stall to $158,000. Equipment costs decreased from $10,000/stall for a double-six parlor to $9000/stall for a double-ten parlor with a cost of $1200/stall for pipeline milking. Initial parlor cost was increased $5000/stall to represent a high cost parlor. Whether it is economically beneficial to invest in an AMS instead of a conventional milking parlor depends on constructions costs, investments in the milking system and costs of labour. Besides costs of labour, the availability of labour should also be taken into account. In general, an AMS is economically beneficial for smaller scale farms, and large dairies can usually operate more cheaply with a milking parlor.

- Increased electricity costs - to operate the robots, but this can be more than outweighed by reduced labour input.

- Increased complexity – While complexity of equipment is a necessary part of technological advancement, the increased complexity of the AMS milking unit over conventional systems, increases the reliance on manufacturer maintenance services and possibly increasing operating costs. The farmer is exposed in the event of total system failure, relying on prompt response from the service provider. In practice AMS systems have proved robust and manufacturers provide good service networks. Because all milking cows have to visit the AMS voluntarily, the system requires a high quality of management. The system also involves a central place for the computer in the daily working routines.

- Difficult to apply in pasture systems – AMS works best in zero-grazing systems, in which the cow is housed indoors for most of the lactation period. Zero-grazing suits areas (e.g. the Netherlands) where land is at a premium, as maximum land can be devoted to feed production which is then collected by the farmer and brought to the animals in the barn. In pasture systems, cows graze in fields and are required to walk to the milking parlour. It has been found that cows tend not to attend the milking unit if the distance to walk is too great. There are currently research projects at the Dexcel facility in New Zealand, University of Sydney's FutureDairy site, and Michigan State University's Kellogg Biological Station, where cattle are on pasture and milked by AMS.

- Lower milk quality – Somatic cell count (SCC) and Plate loop count (PLC) are, respectively, measurements of the quantity of white blood cells and total number of bacteria present in a milk sample. A high SCC indicates reduced udder health (as the immune system fights some infection) and implies lower milk quality. AMS herds consistently show higher SCCs than conventionally milked herds. A high PLC indicates bacterial contamination, usually through poor sanitation or cooling and similarly implies low milk quality. High PLC in AMS may be attributed to the continuous use of milking lines (rather than twice a day in conventional systems), which reduces the time window for cleaning, and the incremental addition of milk to the bulk milk tank which may not cool efficiently at low milk levels.

- Possible increase in stress for some cows – Cows are social animals, and it has been found that due to dominance of some cows, others will be forced to milk only at night. Such behaviour is inconsistent with the perception that AM reduces stress by allowing "free choice" of milking time.

- Decreased contact between farmer and herd – Effective animal husbandry requires that the farmer be fully aware of herd condition. In conventional milking, the cows are observed before milking equipment is attached, and ill or injured cows can be earmarked for attention. Automatic milking removes the farmer from such close contact with the animal, with the possibility that illness may go unnoticed for longer periods and both milk quality and cow welfare suffer. In practice, milk quality sensors at the milking unit attempt to detect changes in milk due to infection, and farmers inspect the herd frequently. However this concern has meant that farmers are still tied to a seven-day schedule. Modern automatic milking systems attempt to rectify this problem by gathering data that would not be available in many conventional systems including milk temperature, milk conductivity, milk color including infrared scan, change in milking speed, change in milking time or milk letdown by quarter, cow's weight, cow's activity (movements), time spent ruminating, etc.

Manufacturers

A DeLaval *VMS* unit, 2007

- Lely (Netherlands), *Lely Astronaut AMS*

- DeLaval (Sweden), *DeLaval VMS*

- Fullwood (UK), *Merlin AMS*

- GEA Farm Technologies (Germany, formerly WestfaliaSurge), *MIone AMS*

- SAC (Denmark), purchased the Dutch manufacturer of the *Galaxy Robot AMS* in 2005, sell under the brands *SAC RDS Futureline MARK II, Insentec Galaxy Starline, BouMatic's ProFlex*

- BoumaticRobotics (NL), *MR-S1, MR-D1*

Yamaha R-MAX

The Yamaha R-MAX is a Japanese unmanned helicopter developed by the Yamaha Motor Company in the 1990s. The two-bladed, gasoline-powered aircraft is remote-controlled by a line-of-sight user; it was designed primarily for agricultural use, and is capable of precise aerial spraying of crops. The R-MAX has been used in Japan and abroad for agriculture and a variety of other roles, including reconnaissance, disaster response and technology development.

Development

The Yamaha R-MAX and its predecessor, the Yamaha R-50, were developed in the 1990s in response to demand in the Japanese market for aerial agricultural spraying. Fixed-wing manned crop dusters had been in use in Japan for many years, but the small size of most Japanese farms meant that this method was inefficient. Manned helicopters were sometimes used for spraying, but were very expensive. The R-MAX allowed much more precise small-scale spraying, at a lower cost and lower risk than manned aircraft. The R-MAX was approved for operation in the United States by the Federal Aviation Administration in 2015.

Operational History

As of 2015, the R-MAX fleet has conducted over two million flight hours in agricultural roles and several other capacities, including aerial sensing, photography, academic research, and military applications.

Volcano Observation

In the spring of 2000, the Japanese government requested the use of an R-MAX to observe the eruption of Mount Usu, which had been dormant for 22 years, as close observation of the volcano was deemed too dangerous for manned helicopters. The R-MAX allowed scientific observers to spot and measure build-ups of volcanic ash which would have otherwise been missed, and improved the observers' ability to predict hazardous volcanic mudslides.

Fukushima Nuclear Disaster

Yamaha R-MAXs were used in the wake of the 2011 Tōhoku earthquake and tsunami to monitor radiation levels around the site of the Fukushima nuclear disaster from inside the "no-entry" zone.

Research

The R-MAX has been used by several universities worldwide for guidance and automatic control research. In 2002, Georgia Tech's UAV Research Facility began using a Yamaha R-MAX for research into autonomous aerial guidance, navigation, and control systems. Carnegie Mellon University, the University of California Berkeley, UC Davis and Virginia Tech have also used R-MAX units for research.

Variants

In May 2014, Yamaha agreed to partner with the American defense firm Northrop Grumman to produce a variant of the R-MAX, known as the R-Bat, combining the R-MAX airframe with a suite of autonomy-enabling hardware and software. This product is intended for both military and civilian applications.

References

- "Self-driving Ibex robot sprayer helps farmers safely tackle hills - Farmers Weekly". Farmers Weekly. Retrieved 2016-03-22.

- Belton, Padraig (2016-11-25). "In the future, will farming be fully automated?". BBC News. Retrieved 2016-11-28.

- Head, Elan (6 May 2015). "FAA grants exemption to unmanned Yamaha RMAX helicopter". Vertical Magazine. Retrieved 6 May 2015.

- "Virginia Tech engineering team developing helicopter that would investigate nuclear disasters". Virginia Tech College of Engineering. 23 March 2010. Retrieved 22 August 2015.

- "Photo Release -- Northrop Grumman, Yamaha Motor, U.S.A., Collaborate on Unmanned Helicopter System". Northrop Grumman via GlobeNewsWire.com. 8 May 2014. Retrieved 12 August 2015.

- Evan Ackerman. "Yamaha Demos Agricultural RoboCopter, But Humans Can't Unleash It Yet". IEEE Spectrum. 16 October 2014. Retrieved 19 August 2015.

- Harvey, Fiona. "Robot farmers are the future of agriculture, says government". The Guardian. Retrieved 30 October 2014.

- Jenkins, David (23 September 2013). "Agriculture shock: How robot farmers will take over our fields". Metro. Retrieved 30 October 2014.

- Foglia, Mario; Reina, Giulio. "Agricultural Robot for Radicchio Harvesting" (PDF). Journal of Field Robotics. Wiley InterScience. doi:10.1002/rob. Retrieved 30 October 2014.

- Monta, M.; Kondo, N.; Shibano, Y. "Agricultural Robot in Grape Production System". Institute of Electrical and Electronics Engineers Xplore Digital Library. Retrieved 30 October 2014.

- "Yamaha Motors Monthly Newsletter" (PDF). Yamaha Motor Company. 18 November 2013. Retrieved 31 October 2014.

Various Terminology used in Agricultural Engineering

Monoculture is the practice of growing a particular crop at a time. It is criticized by many for harming the environment and for putting the food supply chain at risk. The other various terminologies used in agricultural engineering are dosing, fertigation, convertible husbandry, monocropping, windrow and foodshed. The chapter serves as a source to understand the major terms used in agricultural engineering.

Monoculture

Monoculture is the agricultural practice of producing or growing a single crop, plant, or livestock species, variety, or breed in a field or farming system at a time. Polyculture, where more than one crop is grown in the same space at the same time, is the alternative to monoculture. Monoculture is widely used in both industrial farming and organic farming and has allowed increased efficiency in planting and harvest.

Continuous monoculture, or monocropping, where the same species is grown year after year, can lead to the quicker buildup of pests and diseases, and then rapid spread where a uniform crop is susceptible to a pathogen. The practice has been criticized for its environmental effects and for putting the food supply chain at risk. Diversity can be added both in time, as with a crop rotation or sequence, or in space, with a polyculture.

Diversity of crops in space and time; monocultures and polycultures, and rotations of both.					
			Diversity in time		
		Higher			
	Low	Cyclic	Dynamic (non-cyclic)		
Diversity in space	Low	Monoculture, one species in a field	Continuous monoculture, monocropping	Crop rotation (rotation of monocultures)	Sequence of mono-cultures
	Higher	Polyculture, two or more species intermingled in a field	Continuous polyculture	Rotation of polycultures	Sequence of poly-cultures

Oligoculture has been suggested to describe a crop rotation of just a few crops, as is practiced by several regions of the world.

The term monoculture is frequently applied for other uses to describe any group dominated by a single variety, e.g. social Monoculturalism, or in the field of musicology to describe the dominance of the American and British music-industries in Western pop music, or in the field of computer science to describe a group of computers all running identical software.

A monocultivated potato field

Land Use

The term is used in agriculture and describes the practice of planting the same cultivar over an extended area. Each cultivar has the same standardized planting, maintenance and harvesting requirements resulting in greater yields and lower costs.

It also is beneficial because a crop can be tailor-planted for a location that has special problems – like soil salt or drought or a short growing season.

When a crop is matched to its well-managed environment, a monoculture can produce higher yields than a polyculture. In the last 40 years, modern practices such as monoculture planting and the use of synthesized fertilizers have reduced the amount of additional land needed to produce food. However, planting the same crop in the same place each year depletes the nutrients from the earth that the plant relies on and leaves soil weak and unable to support healthy plant growth. Because soil structure and quality is so poor, farmers are forced to use chemical fertilizers to encourage plant growth and fruit production. These fertilizers, in turn, disrupt the natural makeup of the soil and contribute further to nutrient depletion. Monocropping also creates the spread of pests and diseases, which have to be treated with yet more chemicals. The effects of monocropping on the environment are severe when pesticides and fertilizers make their way into ground water or become airborne, creating pollution.

Forestry

In forestry, monoculture refers to the planting of one species of tree. Monoculture plantings provide great yields and more efficient harvesting than natural stands of trees. Single-species stands of trees are often the natural way trees grow, but the stands show a diversity in tree sizes, with dead trees mixed with mature and young trees. In forestry, monoculture stands that are planted and harvested as a unit provide limited resources for wildlife that depend on dead trees and openings, since all the trees are the same size; they are most often harvested by clearcutting, which drastically alters the habitat. The mechanical harvesting of trees can compact soils, which can adversely affect understory growth. Single-species planting of trees also are more vulnerable when infected with a pathogen, or are attacked by insects, and by adverse environmental conditions.

Lawns and Animals

Examples of monoculture include lawns and most field cs wheat or corn. The term is also used where a single breed of farm animal is raised in large-scale concentrated animal feeding operations (CAFOs). In the United States, The Livestock Conservancy was formed to protect nearly 200 endangered livestock breeds from going extinct, largely due to the increased reliance on just a handful of highly specialized breeds.

Disease

Crops used in agriculture are usually single strains that have been bred for high yield and resistant to certain common diseases. Since all plants in a monoculture are genetically similar, if a disease strikes to which they have no resistance, it can destroy entire populations of crops. Polyculture, which is the mixing of different crops, reduces the likelihood that one or more of the crops will be resistant to any particular pathogen. Studies have shown planting a mixture of crop strains in the same field to be effective at combating disease. Ending monocultures grown under disease conditions by introducing crop diversity has greatly increased yields. In one study in China, the planting of several varieties of rice in the same field increased yields of non-resistant strains by 89% compared to non-resistant strains grown in monoculture, largely because of a dramatic (94%) decrease in the incidence of disease, making pesticides less necessary. There is currently a great deal of international worry about the wheat leaf rust fungus, that has already decimated wheat crops in Uganda and Kenya, and is starting to make inroads into Asia as well. As much of the world's wheat crops are very genetically similar following the Green Revolution, the impacts of such diseases threaten agricultural production worldwide.

In Ireland, exclusive use of one variety of potato, the "lumper", led to the great famine. It was inexpensive food to feed the masses. Potatoes were propagated vegetatively with little to no genetic variation. When *Phytophthora infestans* arrived from the Americas in 1845 to Ireland, the lumper had no resistance to the disease leading to the nearly complete failure of the potato crop across Ireland. Had the farmers used multiple varieties of potato, the famine may not have occurred. Andean natives were cultivating three thousand varieties before the Spaniards arrived.

Many of today's livestock production systems rely on just a handful of highly specialized breeds. By focusing heavily on a single trait (output), other traits like fertility, disease resistance, vigor, and mothering instincts are sacrificed. In the early 1990s a few
Holstein calves were observed to grow poorly and died in the first 6 months of life.
They were all found to be homozygous for a mutation in the gene that caused Bovine
Leukocyte Adhesion Deficiency. This mutation was found at a high frequency in Holstein populations world-wide. (15% among bulls in the US, 10% in Germany, and 16%
in Japan.) By studying the pedigrees of affected and carrier animals the source of the
mutation was tracked to a single bull that was widely used in the industry. Note that in
1990 there were approximately 4 million Holteins in the US making the affected population around 600,000 animals.

Polyculture

The environmental movement seeks to change popular culture by redefining the "perfect lawn" to be something other than a turf monoculture, and seeks agricultural policy
that provides greater encouragement for more diverse cropping systems. Local food
systems may also encourage growing multiple species and a wide variety of crops at the
same time and same place. Heirloom gardening and raising heritage livestock breeds
have come about largely as a reaction against monocultures in agriculture.

Tillage

Cultivating after an early rain

Tillage is the agricultural preparation of soil by mechanical agitation of various types,
such as digging, stirring, and overturning. Examples of human-powered tilling methods using hand tools include shovelling, picking, mattock work, hoeing, and raking. Examples of draft-animal-powered or mechanized work include ploughing (overturning
with moldboards or chiseling with chisel shanks), rototilling, rolling with cultipackers
or other rollers, harrowing, and cultivating with cultivator shanks (teeth). Small-scale
gardening and farming, for household food production or small business production,

tends to use the smaller-scale methods above, whereas medium- to large-scale farming tends to use the larger-scale methods. There is a fluid continuum, however. Any type of gardening or farming, but especially larger-scale commercial types, may also use low-till or no-till methods as well.

Tillage is often classified into two types, primary and secondary. There is no strict boundary between them so much as a loose distinction between tillage that is deeper and more thorough (primary) and tillage that is shallower and sometimes more selective of location (secondary). Primary tillage such as ploughing tends to produce a rough surface finish, whereas secondary tillage tends to produce a smoother surface finish, such as that required to make a good seedbed for many crops. Harrowing and rototilling often combine primary and secondary tillage into one operation.

"Tillage" can also mean the land that is tilled. The word "cultivation" has several senses that overlap substantially with those of "tillage". In a general context, both can refer to agriculture. Within agriculture, both can refer to any of the kinds of soil agitation described above. Additionally, "cultivation" or "cultivating" may refer to an even narrower sense of shallow, selective secondary tillage of row crop fields that kills weeds while sparing the crop plants.

Tillage Systems

Reduced Tillage

Plough tilling the field

Reduced tillage leaves between 15 and 30% residue cover on the soil or 500 to 1000 pounds per acre (560 to 1100 kg/ha) of small grain residue during the critical erosion period. This may involve the use of a chisel plow, field cultivators, or other implements. See the general comments below to see how they can affect the amount of residue.

Intensive Tillage

Intensive tillage leaves less than 15% crop residue cover or less than 500 pounds per acre (560 kg/ha) of small grain residue. This type of tillage is often referred to as conventional tillage but as conservational tillage is now more widely used than intensive tillage (in the United States), it is often not appropriate to refer to this type of tillage as conventional. Intensive tillage often involves multiple operations with implements such as a mold board, disk, and/or chisel plow. Then a finisher with a harrow, rolling basket, and cutter can be used to prepare the seed bed. There are many variations.

Conservation Tillage

Conservation tillage leaves at least 30% of crop residue on the soil surface, or at least 1,000 lb/ac (1,100 kg/ha) of small grain residue on the surface during the critical soil erosion period. This slows water movement, which reduces the amount of soil erosion. Conservation tillage also benefits farmers by reducing fuel consumption and soil compaction. By reducing the number of times the farmer travels over the field, farmers realize significant savings in fuel and labor. In most years since 1997, conservation tillage was used in US cropland more than intensive or reduced tillage.

However, conservation tillage delays warming of the soil due to the reduction of dark earth exposure to the warmth of the spring sun, thus delaying the planting of the next year's spring crop of corn.

- No-till - Never use a plow, disk, etc. ever again. Aims for 100% ground cover.

- Strip-Till - Narrow strips are tilled where seeds will be planted, leaving the soil in between the rows untilled.

- Mulch-till

- Rotational Tillage - Tilling the soil every two years or less often (every other year, or every third year, etc.).

- Ridge-Till

- Zone tillage - This form of conservation tillage is further explained below.

Zone Tillage

Zone tillage is a form of modified deep tillage in which only narrow strips are tilled, leaving soil in between the rows untilled. This type of tillage agitates the soil to help reduce soil compaction problems and to improve internal soil drainage.

Purpose

Zone tillage is designed to only disrupt the soil in a narrow strip directly below the crop row. In comparison to no-till, which relies on the previous year's plant residue to protect the soil and aides in postponement of the warming of the soil and crop growth in Northern climates, zone tillage creates approximately a 5-inch-wide strip that simultaneously breaks up plow pans, assists in warming the soil and helps to prepare a seedbed. When combined with cover crops, zone tillage helps replace lost organic matter, slows the deterioration of the soil, improves soil drainage, increases soil water and nutrient holding capacity, and allows necessary soil organisms to survive.

Usage

It has been successfully used on farms in the mid-west and west for over 40 years and is currently used on more than 36% of the U.S. farmland. Some specific states where zone tillage is currently in practice are Pennsylvania, Connecticut, Minnesota, Indiana, Wisconsin, and Illinois.

Unfortunately, there aren't consistent yield results in the Northern Cornbelt states; however, there is still interest in deep tillage within the agriculture industry. In areas that are not well-drained, deep tillage may be used as an alternative to installing more expensive tile drainage.

Effects of Tillage

Positive

Plowing:

- Loosens and aerates the top layer of soil, which facilitates planting the crop
- Helps mix harvest residue, organic matter (humus), and nutrients evenly into the soil
- Mechanically destroys weeds
- Dries the soil before seeding (in wetter climates tillage aids in keeping the soil drier)
- When done in autumn, helps exposed soil crumble over winter through frosting and defrosting, which helps prepare a smooth surface for spring planting

Negative

- Dries the soil before seeding
- Soil loses a lot of nutrients, like nitrogen and fertilizer, and its ability to store water
- Decreases the water infiltration rate of soil. (Results in more runoff and erosion since the soil absorbs water slower than before)
- Tilling the soil results in dislodging the cohesiveness of the soil particles thereby inducing erosion.
- Chemical runoff

- Reduces organic matter in the soil

- Reduces microbes, earthworms, ants, etc.

- Destroys soil aggregates

- Compaction of the soil, also known as a tillage pan

- Eutrophication (nutrient runoff into a body of water)

- Can attract slugs, cut worms, army worms, and harmful insects to the left over residues.

- Crop diseases can be harbored in surface residues

General Comments

- The type of implement makes the most difference, although other factors can have an effect.

- Tilling in absolute darkness (night tillage) might reduce the number of weeds that sprout following the tilling operation by half. Light is necessary to break the dormancy of some weed species' seed, so if fewer seeds are exposed to light during the tilling process, fewer will sprout. This may help reduce the amount of herbicides needed for weed control.

- Greater speeds, when using certain tillage implements (disks and chisel plows), lead to more intensive tillage (i.e., less residue is on the soil surface).

- Increasing the angle of disks causes residues to be buried more deeply. Increasing their concavity makes them more aggressive.

- Chisel plows can have spikes or sweeps. Spikes are more aggressive.

- Percentage residue is used to compare tillage systems because the amount of crop residue affects the soil loss due to erosion.

- See Soybean management practices to see what types of tillage are currently recommended for Soybean Production.

Definitions

Primary tillage loosens the soil and mixes in fertilizer and/or plant material, resulting in soil with a rough texture.

Secondary tillage produces finer soil and sometimes shapes the rows, preparing the seed bed. It also provides weed control throughout the growing season during the maturation of the crop plants, unless such weed control is instead achieved with low-till or no-till methods involving herbicides.

- The seed bed preparation can be done with harrows (of which there are many types and subtypes), dibbles, hoes, shovels, rotary tillers, subsoilers, ridge- or bed-forming tillers, rollers, or cultivators.

- The weed control, to the extent that it is done via tillage, is usually achieved with cultivators or hoes, which disturb the top few centimeters of soil around the crop plants but with minimal disturbance of the crop plants themselves. The tillage kills the weeds via 2 mechanisms: uprooting them, burying their leaves (cutting off their photosynthesis), or a combination of both. Weed control both prevents the crop plants from being outcompeted by the weeds (for water and sunlight) and prevents the weeds from reaching their seed stage, thus reducing future weed population aggressiveness.

History of Tilling

Tilling with Hungarian Grey cattles

Tilling was first performed via human labor, sometimes involving slaves. Hoofed animals could also be used to till soil via trampling. The wooden plow was then invented. It could be pulled by mule, ox, elephant, water buffalo, or similar sturdy animal. Horses are generally unsuitable, though breeds such as the scyne could work. The steel plow allowed farming in the American Midwest, where tough prairie grasses and rocks caused trouble. Soon after 1900, the farm tractor was introduced, which eventually made modern large-scale agriculture possible.

Alternatives to Tilling

Modern agricultural science has greatly reduced the use of tillage. Crops can be grown for several years without any tillage through the use of herbicides to control weeds, crop varieties that tolerate packed soil, and equipment that can plant seeds or fumigate the soil without really digging it up. This practice, called no-till farming, reduces costs and environmental change by reducing soil erosion and diesel fuel usage.

Site Preparation of Forest Land

Site preparation is any of various treatments applied to a site in order to ready it for seeding or planting. The purpose is to facilitate the regeneration of that site by the chosen method. Site preparation may be designed to achieve, singly or in any combination: improved access, by reducing or rearranging slash, and amelioration of adverse forest floor, soil, vegetation, or other biotic factors. Site preparation is undertaken to ameliorate one or more constraints that would otherwise be likely to thwart the objectives of management. A valuable bibliography on the effects of soil temperature and site preparation on subalpine and boreal tree species has been prepared by McKinnon et al. (2002).

Site preparation is the work that is done before a forest area is regenerated. Some types of site preparation are burning.

Burning

Broadcast burning is commonly used to prepare clearcut sites for planting, e.g., in central British Columbia, and in the temperate region of North America generally.

Prescribed burning is carried out primarily for slash hazard reduction and to improve site conditions for regeneration; all or some of the following benefits may accrue:

> a) Reduction of logging slash, plant competition, and humus prior to direct seeding, planting, scarifying or in anticipation of natural seeding in partially cut stands or in connection with seed-tree systems.

> b) Reduction or elimination of unwanted forest cover prior to planting or seeding, or prior to preliminary scarification thereto.

> c) Reduction of humus on cold, moist sites to favour regeneration.

> d) Reduction or elimination of slash, grass, or brush fuels from strategic areas around forested land to reduce the chances of damage by wildfire.

Prescribed burning for preparing sites for direct seeding was tried on a few occasions in Ontario, but none of the burns was hot enough to produce a seedbed that was adequate without supplementary mechanical site preparation.

Changes in soil chemical properties associated with burning include significantly increased pH, which Macadam (1987) in the Sub-boreal Spruce Zone of central British Columbia found persisting more than a year after the burn. Average fuel consumption was 20 to 24 t/ha and the forest floor depth was reduced by 28% to 36%. The increases correlated well with the amounts of slash (both total and ≥7 cm diameter) consumed. The change in pH depends on the severity of the burn and the amount consumed; the increase can be as much as 2 units, a 100-fold change. Deficiencies of copper and iron in the foliage of white spruce on burned clearcuts in central British Columbia might be attributable to elevated pH levels.

Even a broadcast slash fire in a clearcut does not give a uniform burn over the whole area. Tarrant (1954), for instance, found only 4% of a 140-ha slash burn had burned severely, 47% had burned lightly, and 49% was unburned. Burning after windrowing obviously accentuates the subsequent heterogeneity.

Marked increases in exchangeable calcium also correlated with the amount of slash at least 7 cm in diameter consumed. Phosphorus availability also increased, both in the forest floor and in the 0 cm to 15 cm mineral soil layer, and the increase was still evident, albeit somewhat diminished, 21 months after burning. However, in another study in the same Sub-boreal Spruce Zone found that although it increased immediately after the burn, phosphorus availability had dropped to below pre-burn levels within 9 months.

Nitrogen will be lost from the site by burning, though concentrations in remaining forest floor were found by Macadam (1987) to have increased in 2 of 6 plots, the others showing decreases. Nutrient losses may be outweighed, at least in the short term, by improved soil microclimate through the reduced thickness of forest floor where low soil temperatures are a limiting factor.

The *Picea/Abies* forests of the Alberta foothills are often characterized by deep accumulations of organic matter on the soil surface and cold soil temperatures, both of which make reforestation difficult and result in a general deterioration in site productivity; Endean and Johnstone (1974) describe experiments to test prescribed burning as a means of seedbed preparation and site amelioration on representative clear-felled *Picea/Abies* areas. Results showed that, in general, prescribed burning did not reduce organic layers satisfactorily, nor did it increase soil temperature, on the sites tested. Increases in seedling establishment, survival, and growth on the burned sites were probably the result of slight reductions in the depth of the organic layer, minor increases in soil temperature, and marked improvements in the efficiency of the planting crews. Results also suggested that the process of site deterioration has not been reversed by the burning treatments applied.

Ameliorative Intervention

Slash weight (the oven-dry weight of the entire crown and that portion of the stem < 4 inches in diameter) and size distribution are major factors influencing the forest fire hazard on harvested sites. Forest managers interested in the application of prescribed burning for hazard reduction and silviculture, were shown a method for quantifying the slash load by Kiil (1968). In west-central Alberta, he felled, measured, and weighed 60 white spruce, graphed (a) slash weight per merchantable unit volume against diameter at breast height (dbh), and (b) weight of fine slash (<1.27 cm) also against dbh, and produced a table of slash weight and size distribution on one acre of a hypothetical stand of white spruce. When the diameter distribution of a stand is unknown, an estimate of slash weight and size distribution can be obtained from average stand diameter, number of trees per unit area, and merchantable cubic foot volume. The sample trees in Kiil's study had full symmetrical crowns. Densely growing trees with short and often irregular crowns would probably be overestimated; open-grown trees with long crowns would probably be underestimated.

The need to provide shade for young outplants of Engelmann spruce in the high Rocky Mountains is emphasized by the U.S. Forest Service. Acceptable planting spots are defined as microsites on the north and east sides of down logs, stumps, or slash, and lying in the shadow cast by such material. Where the objectives of management specify more uniform spacing, or higher densities, than obtainable from an existing distribution of shade-providing material, redistribution or importing of such material has been undertaken.

Access

Site preparation on some sites might be done simply to facilitate access by planters, or to improve access and increase the number or distribution of microsites suitable for planting or seeding.

Wang et al. (2000) determined field performance of white and black spruces 8 and 9 years after outplanting on boreal mixedwood sites following site preparation (Donaren disc trenching versus no trenching) in 2 plantation types (open versus sheltered) in southeastern Manitoba. Donaren trenching slightly reduced the mortality of black spruce but significantly increased the mortality of white spruce. Significant difference in height was found between open and sheltered plantations for black spruce but not for white spruce, and root collar diameter in sheltered plantations was significantly larger than in open plantations for black spruce but not for white spruce. Black spruce open plantation had significantly smaller volume (97 cm^3) compared with black spruce sheltered (210 cm^3), as well as white spruce open (175 cm^3) and sheltered (229 cm^3) plantations. White spruce open plantations also had smaller volume than white spruce sheltered plantations. For transplant stock, strip plantations had a significantly higher volume (329 cm^3) than open plantations (204 cm^3). Wang et al. (2000) recommended that sheltered plantation site preparation should be used.

Mechanical

Up to 1970, no "sophisticated" site preparation equipment had become operational in Ontario, but the need for more efficacious and versatile equipment was increasingly recognized. By this time, improvements were being made to equipment originally developed by field staff, and field testing of equipment from other sources was increasing.

According to J. Hall (1970), in Ontario at least, the most widely used site preparation technique was post-harvest mechanical scarification by equipment front-mounted on a bulldozer (blade, rake, V-plow, or teeth), or dragged behind a tractor (Imsett or S.F.I. scarifier, or rolling chopper). Drag type units designed and constructed by Ontario's Department of Lands and Forests used anchor chain or tractor pads separately or in combination, or were finned steel drums or barrels of various sizes and used in sets alone or combined with tractor pad or anchor chain units.

J. Hall's (1970) report on the state of site preparation in Ontario noted that blades and rakes were found to be well suited to post-cut scarification in tolerant hardwood stands for natural regeneration of yellow birch. Plows were most effective for treating dense brush prior to planting, often in conjunction with a planting machine. Scarifying teeth, e.g., Young's teeth, were sometimes used to prepare sites for planting, but their most effective use was found to be preparing sites for seeding, particularly in backlog areas carrying light brush and dense herbaceous growth. Rolling choppers found application in treating heavy brush but could be used only on stone-free soils. Finned drums were commonly used on jack pine–spruce cutovers on fresh brushy sites with a deep duff layer and heavy slash, and they needed to be teamed with a tractor pad unit to secure good distribution of the slash. The S.F.I. scarifier, after strengthening, had been "quite successful" for 2 years, promising trials were under way with the cone scarifier and barrel ring scarifier, and development had begun on a new flail scarifier for use on sites with shallow, rocky soils. Recognition of the need to become more effective and efficient in site preparation led the Ontario Department of Lands and Forests to adopt the policy of seeking and obtaining for field testing new equipment from Scandinavia and elsewhere that seemed to hold promise for Ontario conditions, primarily in the north. Thus, testing was begun of the Brackekultivator from Sweden and the Vako-Visko rotary furrower from Finland.

Mounding

Site preparation treatments that create raised planting spots have commonly improved outplant performance on sites subject to low soil temperature and excess soil moisture. Mounding can certainly have a big influence on soil temperature. Draper et al. (1985), for instance, documented this as well as the effect it had on root growth of outplants (Table 30).

The mounds warmed up quickest, and at soil depths of 0.5 cm and 10 cm averaged 10 and 7 °C higher, respectively, than in the control. On sunny days, daytime surface temperature maxima on the mound and organic mat reached 25 °C to 60 °C, depending on soil wetness and shading. Mounds reached mean soil temperatures of 10 °C at 10 cm depth 5 days after planting, but the control did not reach that temperature until 58 days after planting. During the first growing season, mounds had 3 times as many days with a mean soil temperature greater than 10 °C than did the control microsites.

Draper et al.'s (1985) mounds received 5 times the amount of photosynthetically active radiation (PAR) summed over all sampled microsites throughout the first growing season; the control treatment consistently received about 14% of daily background PAR, while mounds received over 70%. By November, fall frosts had reduced shading, eliminating the differential. Quite apart from its effect on temperature, incident radiation is also important photosynthetically. The average control microsite was exposed to levels of light above the compensation point for only 3 hours, i.e., one-quarter of the daily light period, whereas mounds received light above the compensation point for 11 hours, i.e., 86% of the same daily period. Assuming that incident light in the 100-600 $\mu Em^{-2}s^{-1}$ intensity range is the most important for photosynthesis, the mounds received over 4 times the total daily light energy that reached the control microsites.

Orientation of Linear Site Preparation, e.g., Disk-trenching

With linear site preparation, orientation is sometimes dictated by topography or other considerations, but the orientation can often be chosen. It can make a difference. A disk-trenching experiment in the Sub-boreal Spruce Zone in interior British Columbia investigated the effect on growth of young outplants (lodgepole pine) in 13 microsite planting positions: berm, hinge, and trench in each of north, south, east, and west aspects, as well as in untreated locations between the furrows. Tenth-year stem volumes of trees on south, east, and west-facing microsites were significantly greater than those of trees on north-facing and untreated microsites. However, planting spot selection was seen to be more important overall than trench orientation.

In a Minnesota study, the N–S strips accumulated more snow but snow melted faster than on E–W strips in the first year after felling. Snow-melt was faster on strips near the centre of the strip-felled area than on border strips adjoining the intact stand. The strips, 50 feet (15.24 m) wide, alternating with uncut strips 16 feet (4.88 m) wide, were felled in a *Pinus resinosa* stand, aged 90 to 100 years.

Anden

Andenes in the Sacred Valley, close to Pisac, Perú

Andenes are terraces dug into the slopes of mountains for agricultural purposes. They were constructed and much used in the Andes mountain range to provide cultivable hillsides. The majority of these terraces were constructed and used by the pre-Hispanic cultures, and many can still be observed throughout the region.

Origin and History

The rivers that flow through the Andean mountain range form narrow valleys in the regions above altitudes of 500 meters. Unlike the topography of the Peruvian coast where irrigation by canals allows cultivation of the desert plains, the very narrow and deep valleys in the mountainous zones prohibit large scale agriculture. The ancient Andeans, who needed farmland in addition to that provided by their narrow valleys, attempted to gain more usable land at the cost of the mountains and created the first andenes.

The scale of the andenes does not seem to have been very important until approximately the sixth century, when the Wari, or Huari, government began a mass construction of Andenes in the region of Ayacucho, which involved a large inversion of the labor force. At this time the Wari acquired geopolitical importance and began their expansion into the Central Andes in what is considered the first Andean empire (500–900 AD).

In the successive centuries they refined the technique of construction of the Andenes, incorporating layers of different materials into the filling, in order to better control drainage compared to the same rainfall. In the fifteenth century the Incas brought the architecture of the Andenes to their utmost splendor, investing considerable resources not only in the filling but also in the quality of the stone walls.

In the Incan period the Andenes were used to other ends, specifically to control the erosion of the mountains where they constructed their ceremonial centers. For example, a good part of Andean construction in the extreme west of Machu Picchu appears to be structural.

After the Spanish conquest, the andenes continued in use. They have been maintained to this day, and still account for a significant amount of cultivation.

Famous Collections of Andenes

The Andenes possess an appeal beyond the historical and their original economic motivations: they are also landscape resources whose situation in the Andes Mountains has notable aesthetic value. Many of them follow the natural curve of the slopes in such a way that preserves the visual harmony of the environment. The idea of hanging gardens in the mountains can fit well with the description of the Andenes.

Between the center of Peru and the north of Bolivia one finds the best conserved collection of andenes. Perhaps the most impressive Andenes zone is the Colca Canyon (Valle de Colca), whose terraces were constructed by the Collaguas beginning in the 11th century. Those on the islands in Lake Titicaca (constructed by the Aymara) are visually stunning, as are those in the so-called Sacred Valley of the Incas (Valle Sagrado de los Incas) in Cusco, those constructed by the Incas in Moray (Inca ruin) in a collection of concentric circles, as well as the enormous terraces at Pisaq and Ollantaytambo. A good part of these Andenes are used to this day, which illustrates the quality of their design.

Dosing

Dosing generally applies to feeding chemicals or medicines in small quantities into a process fluid or to a living being at intervals or to atmosphere at intervals to give sufficient time for the chemical or medicine to react or show the results.

In the case of human beings or animals the word *dose* is generally used but in the case of inanimate objects the word dosing is used. The term dose titration, referring to stepwise titration of doses until a desired level of effect is reached, is common in medicine.

In Engineering

The word *dosing* is very commonly used by engineers in thermal power stations, in water treatment, in any industry where steam is being generated, and in building services for heating and cooling water treatment. Dosing procedures are also in vogue in textile and similar industries where chemical treatment is involved.

Commercial swimming pools also require chemical dosing in order to control pH balance, chlorine level, and other such water quality criteria. Modern swimming pool plant will have bulk storage of chemicals held in separate dosing tanks, and will have automated controls and dosing pumps to top up the various chemicals as required to control the water quality.

In a power station treatment chemicals are injected or fed to boiler and also to feed and make up water under pressure, but in small dosages or rate of injection. The feeding at all places is done by means of small capacity dosing pumps specially designed for the duty demanded.

In building services the water quality of various pumped fluid systems, including for heating, cooling, and condensate water, will be regularly checked and topped up with chemicals manually as required to suit the required water quality. Most commonly inhibitors will be added to protect the pipework and components against corrosion, or a biocide will be added to stop the growth of bacteria in lower temperature systems. The required chemicals will be added to the fluid system by use of a dosing pot; a multi-valved chamber in which the chemical can be added, and then introduced to the fluid system in a controlled manner.

In Agriculture

The feeding of chemicals in agriculture has also become common due to technology developments. However agricultural dosing is done by means of hand held pressure spray pumps

Aerial Spraying

Sometimes aerial spraying of chemicals by fixed quantities at intervals or dosing is also adopted for agricultural spraying or for atmospheric spraying for eliminating certain types of harmful insects.

Fertigation

Fertigation using white poly bag

Fertigation is the injection of fertilizers, soil amendments, and other water-soluble products into an irrigation system.

Fertigation is related to chemigation, the injection of chemicals into an irrigation system. The two terms are sometimes used interchangeably however chemigation is generally a more controlled and regulated process due to the nature of the chemicals used. Chemigation often involves pesticides, herbicides, and fungicides, some of which pose health threat to humans, animals, and the environment.

1. Numbered list item

2. Numbered list item

Numbered List Item

Uses

Fertigation is practiced extensively in commercial agriculture and horticulture. Fertigation is also increasingly being used for landscaping as dispenser units become more reliable and easier to use. Fertigation is used to add additional nutrients or to correct nutrient deficiencies detected in plant tissue analysis. It is usually practiced on high-value crops such as vegetables, turf, fruit trees, and ornamentals.

Commonly Used Nutrients

Most plant nutrients can be applied through irrigation systems. Nitrogen is the most commonly used plant nutrient. Naturally occurring nitrogen (N_2) is a diatomic molecule which makes up approximately 80% of the earth's atmosphere. Most plants cannot directly consume diatomic nitrogen, therefore nitrogen must be contained as a component of other chemical substances which plants can consume. Commonly, anhydrous ammonia, ammonium nitrate, and urea are used as bioavailable sources of nitrogen. Other nutrients needed by plants include phosphorus and potassium. Like nitrogen, plants require these substances to live but they must be contained in other chemical substances such as monoammonium phosphate or diammonium phosphate to serve as bioavailable nutrients. A common source of potassium is muriate of potash which is chemically potassium chloride. A soil fertility analysis is used to determine which of the more stable nutrients should be used.

Advantages

The benefits of fertigation methods over conventional or drop-fertilizing methods include:

- Increased nutrient absorption by plants.

- Reduction of fertilizer, chemicals, and water needed.

- Reduced leaching of chemicals into the water supply.

- Reduced water consumption due to the plant's increased root mass's ability to trap and hold water.

- Application of nutrients can be controlled at the precise time and rate necessary.

- Minimized risk of the roots contracting soil borne diseases through the contaminated soil.

- Reduction of soil erosion issues as the nutrients are pumped through the water drip system.

Disadvantages

- Concentration of the solution decreases as the fertilizer dissolves. This may lead to poor nutrient placement.

- The water supply for fertigation is to be kept separate from the domestic water supply to avoid contamination.

- Possible pressure loss in the main irrigation line.

- The process is dependent on the water supply's non-restriction by drought rationing.

Methods Used

- Drip irrigation-Less wasteful than sprinklers.

- Sprinkler systems-Increases leaf and fruit quality.

- Continuous application-Fertilizer is supplied at a constant rate.

- Three-stage application-Irrigation starts without fertilizers. Fertilizers are applied later in the process.

- Proportional application-Injection rate is proportional to water discharge rate.

- Quantitative application-Nutrient solution is applied in a calculated amount to each irrigation block.

- Other methods of application include the lateral move, the traveler gun, and solid set systems.

1. Fiqah Chan#

System Design

Fertigation assists distribution of fertilizers for farmers. The simplest type of fertigation system consists of a tank with a pump, distribution pipes, capillaries, and a dripper pen.

All systems should be placed on a raised or sealed platform, not in direct contact with the earth. Each system should also be fitted with chemical spill trays.

Because of the potential risk of contamination in the potable (drinking) water supply, a backflow prevention device is required for most fertigation systems. Backflow requirements may vary greatly. Therefore, it is very important to understand the proper level of backflow prevention required by law. In the United States, the minimum backflow protection is usually determined by state regulation. Each city or town may set the level of protection required.

Convertible Husbandry

Within agriculture, convertible husbandry, also known as alternate husbandry, ley husbandry or up-and-down husbandry, was a process used during the 16th century through the 19th century by "which a higher proportion of land was used to support increasing numbers of livestock in many parts of England." In the words of historian Eric Kerridge, convertible husbandry consisted of "the floating of water-meadows, the substitution of up-and-down husbandry for permanent tillage and permanent grass or for shifting cultivation, the introduction of new fallow crops and selected grasses, marsh drainage, manuring, and stock breeding." Convertible husbandry is considered one of the most important changes of the agricultural revolution.

Description

Convertible husbandry was a process that consisted of "alternating arable and pasture on a given piece of land...[by doing so]... farmers almost eliminated the need for fallows between their grain crops and were able to control the quality of their pasture by sowing grass seeds." Alternate husbandry has also been praised as the "best way to keep high fertility on both arable and pasture and to retain excellent soil texture and composition." This system utilized fertilizer in the form of animal manure. Fertilizer was used in greater quantities due to the increase in animal husbandry and resulted in benefiting crop yields when it was time for tillage. Convertible husbandry also came with the added benefit of allowing variations in the types of soils and the extent of leys used in rotation because it was a system in which multiple variables could be modified to suit the needs of the location/type of land/type of soil.

Historical Context

Before the 16th and 17th centuries, farmlands had mostly been founded on the idea of simple alternations of tilling and fallowing during different seasons over several years. However, as livestock became an increasing staple in the lives of farmers and society alike in the midlands, the "rising population...density of settlements, lack of wastelands into which cultivation could expand...and 15th century enclosures for sheep" all led to a need for an improved system of agriculture that allowed for increasing numbers of livestock and a greater increase in crop output compared to input. "The introduction of up-and-down husbandry... helped solve the problems of the midland by providing a measured pasture and arable rotation which not only produced the same amount of grain on a much reduced area, but broke the agrarian cycle of diminished returns by allowing more sheep and cattle to be kept, animals whose dung maintained the fertility of the arable." Another factor that increased the popularity of convertible husbandry had to do with the fact that skill levels of workers were changing and the skills needed to manage a permanent grassland system were acquired too slowly to respond adequately to the growing demand of the population at the time.

Although debatable, many agricultural historians also believe that the introduction of convertible husbandry was brought about with the introduction of the turnip. They argue that "the lowly turnip made possible a change in crop rotation which did not require much capital, but which brought about a tremendous rise in agricultural productivity." They believe that this "fodder" crop pushed agriculture in a direction in which "alternating" husbandry was seen as more efficient than traditional permanent pasture farming and jump-started the improvement of crop rotation and agricultural output versus capital. Although the turnip was popularized by Lord Townshed during the mid-18th century, the use of turnips being grown as fodder was seen as early as the 16th century.

Monocropping

Monocropping is the agricultural practice of growing a single crop year after year on the same land, in the absence of rotation through other crops or growing multiple crops on the same land (polyculture). Corn, soybeans, and wheat are three common crops often grown using monocropping techniques.

While economically a very efficient system, allowing for specialization in equipment and crop production, monocropping is also controversial, as it can damage the soil ecology (including depletion or reduction in diversity of soil nutrients) and provide an unbuffered niche for parasitic species, increasing crop vulnerability to opportunistic insects, plants, and microorganisms. The result is a more fragile ecosystem with an increased dependency on pesticides and artificial fertilizers. The concentrated presence of a single cultivar, genetically adapted with a single resistance strategy, presents a situation in which an entire crop can be wiped out very quickly by a single opportunistic species. An example of this would be the potato famine of Ireland in 1845–1849, and according to Devlin Kuyek is the main cause of the current food crisis with monoculture rice crops failing as the effects of climate change become more acute.

Monocropping as an agricultural strategy tends to emphasize the use of expensive specialized farm equipment — an important component in realizing its efficiency goals. This can lead to an increased dependency on fossil fuels and reliance on expensive machinery that cannot be produced locally and may need to be financed. This can make a significant change in the economics of farming in regions that are accustomed to self-sufficiency in agricultural production. In addition, political complications may ensue when these dependencies extend across national boundaries.

The controversies surrounding monocropping are complex, but traditionally the core issues concern the balance between its advantages in increasing short-term food production — especially in hunger-prone regions — and its disadvantages with respect to long-term land stewardship and the fostering of local economic independence and ecological sustainability. Advocates of monocropping tend to claim that in its absence many human populations would be reduced to starvation or to a degraded level of civilization comparable to the Dark Ages. On the other hand, critics of monocropping dispute these claims and attribute them to corporate special interest groups, citing the damage that monocropping causes to societies and the environment.

A difficulty with monocropping is that the solution to one problem — whether economic, environmental or political — may result in a cascade of other problems. For example, a well-known concern is pesticides and fertilizers seeping into surrounding soil and groundwater from extensive monocropped acreage in the U.S. and abroad. This issue, especially with respect to the pesticide DDT, played an important role in focusing public attention on ecology and pollution issues during the 1960s when Rachel Carson published her landmark book Silent Spring.

Soil depletion is also a negative effect of mono-cropping. Crop rotation plays an important role in replenishing soil nutrients, especially atmospheric nitrogen converted to usable forms by nitrogen-fixing plants used in fallow fields. In addition, it performs an important role in preventing pathogen and pest build-up. In a monocropping regime, farmers are less likely to rotate their crops and replenish such essential soil nutrients. In addition, artificial high-nitrogen fertilizers can "burn" the soil by creating an unfavorable environment for indigenous organisms, a phenomenon well-known to organic gardeners and farmers (who avoid it), resulting in further disruption of soil ecology and dependence on further short-term fertilizer strategies. Lacking a stable ecology, in the absence of substantial irrigation and chemical "fixes" the soil can become dry and begin to erode. As the soil becomes arid and useless, the need for more land becomes an issue, leading to the destruction of even more land — a high-tech version of slash and burn agriculture.

Under certain circumstances monocropping can lead to deforestation (Tauli-Corpuz;Tamang, 2007) or the displacement of indigenous peoples (Tauli-Corpuz;Tamang, 2007).

In order to help reduce dependence on fossil fuels the U.S. government subsidizes the monocropping of corn and soybeans to be used in ethanol production (S, 2007). However monocropping itself is highly chemical- and energy-intensive, as studies by Nelson (2006) indicate. Such studies have shown that the "hidden" energy costs associated with producing each unit of bio-fuel are significantly larger than the amount of energy available from the fuel itself.

Windrow

A windrow is a row of cut (mown) hay or small grain crop. It is allowed to dry before being baled, combined, or rolled. For hay, the windrow is often formed by a hay rake, which rakes hay that has been cut by a mowing machine or by scythe into a row, or it may naturally form as the hay is mown. For small grain crops which are to be harvested, the windrow is formed by a swather which both cuts the crop and forms the windrow.

Windrows of straw, along with stubble.

Grass silage in a windrow

By analogy, the term may also be applied to a row of any other material such as snow, earth or materials for collection. Snow windrows are created by snow plows when clearing roads of snow; where this blocks driveways the windrow may require removal. Snow windrowed to the centre of the street can be removed by a snow blower and truck. In preparing a pond or lake for ice cutting, the snow on top of the ice, which slows freezing, may be scraped off and windrowed.Earth windrows may be formed by graders when grading earthworks or dirt roadsLeaf windrows may be required for municipal collection. Fossil windrows, also 'gyres', are a grouping of fossils that have been deposited together as a result of turbulence or wave action in a marine or freshwater environment. Fossils of similar shape and size are commonly found grouped or sorted together as a result of separation based on weight and shape.Seaweed windrows form on sea or lake surfaces because of cylindrical Langmuir circulation just under the surface caused by wind action. Windrow composting is a large scale vermicomposting system where garden and other biodegradable waste is shredded, mixed and windrowed for composting.

Foodshed

A foodshed is the geographic region that produces the food for a particular population. The term is used to describe a region of food flows, from the area where it is produced, to the place where it is consumed, including: the land it grows on, the route it travels, the markets it passes through, and the tables it ends up on. "Foodshed" is described as a "socio-geographic space: human activity embedded in the natural integument of a particular place." A foodshed is analogous to a watershed in that foodsheds outline the flow of food feeding a particular population, whereas watersheds outline the flow of water draining to a particular location. Through drawing from the conceptual ideas of the watershed, foodsheds are perceived as hybrid social and natural constructs.

It can pertain to the area from which an individual or population receives a particular type of food, or the collective area from which an individual or population receives all of their food. The size of the foodshed can vary depending on the availability of year round foods and the variety of foods grown and processed. Variables such as micro-weather patterns, soil types, water availability, slope conditions, etc. play a role in determining the potential and risk of agriculture).

The modern United States foodshed, as an example, spans the entire world as the foods available in the typical supermarket have traveled from all over the globe, often long distances from where they were produced.

Origin

The term was coined in 1929 in the book *How Great Cities Are Fed* by W.P. Hedden, who was at the time Chief of the Bureau of Commerce for the Port of New York Authority. Hedden described a 'foodshed' in 1929 as the 'dikes and dams' guiding the flow of food from the producer to consumer. Hedden contrasts foodsheds with watersheds by noting that "the barriers which deflect raindrops into one river basin rather than into another are natural land elevations, while the barriers which guide and control movements of foodstuffs are more often economic than physical." Hedden describes the economic forces that influence where foods are produced and how they are transported to the cities in which they are consumed. The term has more recently been reintroduced by permaculturist Arthur Getz, in his 1991 article "Urban Foodsheds" in "Permaculture Activist", to provide an image that helps people to understand how food systems work and that suggests food comes from a source that must be protected.

Foodsheds in American History

Eating within a local foodshed was once the only way in which families gained access to food. In the seventeenth or eighteenth century, most ingredients were drawn from an area of less than fifty acres. There was an interdependence of farming and what was cooked in the kitchen. Farmers gained a sensibility about the land—improved and well-tended land could yield a cornucopian spread and was regarded as a source of food and a sign of wealth. Envisioning and knowing a landscape as one's fount of food is different from what most of us know and experience when driving past fields in the countryside today. People ate food that was in season, when available, or that was preserved.

Very few items came from afar, and if they did they came in small amounts, such as cinnamon and nutmeg. Growing, cooking, and eating food connected most people in preindustrial America to the land.

Methods of Distributing Food Within a Local Foodshed

The "farm-to-table" movement is focused on producing food locally within a foodshed, and delivering it to local consumers. Direct farm-to-table in the United States tends to comprise only a very minor segment of the food distribution system in terms of size and importance, but is growing in popularity.

- Farmers' markets: centerpiece of alternative food distribution systems. The first certified farmers' markets began appearing roughly 25 years ago and are a result of a long-standing desire to protect consumers from fraudulent behavior on the part of resellers. Certified markets are closely regulated by the various state legislators and are required to guarantee that the person selling the produce is actually the person grew the produce.

- Roadside stands: Used by producers to sell fruits and vegetables directly to consumers. These stands help to reduce transportation costs for farmers by bringing the consumer to the produce.)

- Pick-your-own: Farmers open their fields to consumers and allow them to personally select and harvest various types of produce. This method offers the greatest potential savings for both farmers and consumers, because consumers are able to pick produce of the highest quality, and farmers save costs associated with harvesting and marketing.

- Entertainment farming: farmers have been able to use non-revenue-generating farm activities such as walking trails, hay-rides, and animal petting areas to attract consumers.

- Subscription farming: enables consumers to purchase a share of a particular farm's production output. Consumers typically pay a subscription fee for the right to purchase fresh produce during harvest time. Subscribers are then charged for the produce depending on the type and quantity.

- Community-supported agriculture (CSA): allows consumers to purchase shares of a farm's production output.

Nearly 5 percent of all farmers engage in some form of direct food marketing. Estimates of all farm-to-table sales within a foodshed range from roughly $550 million to $2 billion. Many local farms are family-run farms that are successful and do survive through poor economic conditions. Ecologists consider to be more adaptive and more likely to "reproduce" in highly variable and uncertain environments.

Local Foodshed Mapping

The internet can be used to locate foodshed maps of almost any area. Some maps are interactive, where sources in an area can be found for organic produce, microbreweries, farmers' markets, orchards, cheese makers, or other specific categories within a 100-mile radius. A 100- mile radius is considered "local food" because it is large enough to reach beyond a big city, and small enough to feel truly local.

Foodsheds and Sustainability

Buying local food within a foodshed can be seen as a means to combat the modern food system, and the effects it has on the environment. It has been described as "a banner under which people attempt to counteract trends of economic concentration, social disempowerment and environmental degradation in the food and agricultural landscape." Choosing to buy local produce improves the environmental stewardship of producers by reducing the amount of energy used in the transport of foods, as well as greenhouse gas emissions. Agriculture production alone contributes to 14% of anthropogenic greenhouse gas emissions. The food system's contribution of greenhouse gases contributes to the global issue of Climate change. More attention is being paid to possibilities for reducing emissions through more efficient transport and different patterns of consumption, specifically increased reliance on local foodsheds.

One common measure of "eating within a foodshed" is whether produce has traveled under 100 "food miles." Food miles are a measure of how far food travels from the farm where it is produced to the table where it is consumed. In the United States, on average, food travels about 1,500 miles before it gets to a plate. Sale of locally grown food can pave the way for reduction of food miles, and increase in agricultural sustainability.

References

- Richardson, Edited by David M. (2000). Ecology and biogeography of Pinus. Cambridge, U.K. p. 371. ISBN 978-0-521-78910-3.

- Haspel, Tamar (9 May 2014). "Monocrops: They're a problem, but farmers aren't the ones who can solve it.". New York Times. Retrieved 12 January 2016.

- Williams, J.L. (2015-10-22). "The Value of Genome Mapping for the Genetic Conservation of Cattle". The Food and Agriculture Organization of the United Nations. Rome. Retrieved 2015-10-22.

- Mahdi Al-Kaisi; Mark Hanna; Michael Tidman (13 May 2002). "Methods for measuring crop residue". Iowa State University. Retrieved 2012-12-28.

- "Soil Compaction and Conservation Tillage". Conservation Tillage Series. PennState- College of Agricultural Sciences - Cooperative Extension. Retrieved 26 March 2011.

- Vidal, John (2009-03-19). "'Stem rust' fungus threatens global wheat harvest". The Guardian. London. Retrieved 2010-05-13.

Allied Fields of Agricultural Engineering

The allied fields related to agricultural engineering are agricultural chemistry, agricultural diversification, agricultural economics, agricultural philosophy, agroecology, agrophysics etc. Agricultural chemistry is the study of chemistry and biochemistry; they are both equally important for the production of agriculture. This section discusses the allied fields of agricultural engineering in a critical manner providing key analysis to the subject matter.

Agricultural Chemistry

Agricultural chemistry is the study of both chemistry and biochemistry which are important in agricultural production, the processing of raw products into foods and beverages, and in environmental monitoring and remediation. These studies emphasize the relationships between plants, animals and bacteria and their environment. The science of chemical compositions and changes involved in the production, protection, and use of crops and livestock. As a basic science, it embraces, in addition to test-tube chemistry, all the life processes through which humans obtain food and fiber for themselves and feed for their animals. As an applied science or technology, it is directed toward control of those processes to increase yields, improve quality, and reduce costs. One important branch of it, chemurgy, is concerned chiefly with utilization of agricultural products as chemical raw materials.

Sciences

The goals of agricultural chemistry are to expand understanding of the causes and effects of biochemical reactions related to plant and animal growth, to reveal opportunities for controlling those reactions, and to develop chemical products that will provide the desired assistance or control. Every scientific discipline that contributes to agricultural progress depends in some way on chemistry. Hence agricultural chemistry is not a distinct discipline, but a common thread that ties together genetics, physiology, microbiology, entomology, and numerous other sciences that impinge on agriculture.

Chemical materials developed to assist in the production of food, feed, and fiber include scores of herbicides, insecticides, fungicides, and other pesticides, plant growth regulators, fertilizers, and animal feed supplements. Chief among these groups from the commercial point of view are manufactured fertilizers, synthetic pesticides (including herbicides), and supplements for feeds. The latter include both nutritional supplements (for example, mineral nutrients) and medicinal compounds for the prevention or control of disease.

Agricultural chemistry often aims at preserving or increasing the fertility of soil, maintaining or improving the agricultural yield, and improving the quality of the crop.

When agriculture is considered with ecology, the sustainablility of an operation is considered. Modern agrochemical industry has gained a reputation for maximising profits while violating sustainable and ecologically viable agricultural principles. Eutrophication, the prevalence of genetically modified crops and the increasing concentration of chemicals in the food chain (e.g. persistent organic pollutants) are only a few consequences of naive industrial agriculture.

History

- In 1761 Johan Gottschalk Wallerius publishes his pioneering work, *Agriculturae fundamenta chemica* (*Åkerbrukets chemiska grunder*).

- In 1815 Humphry Davy publishes *Elements of agricultural chemistry*

- In 1842 Justus von Liebig publishes *Animal Chemistry or Organic Chemistry in its applications to Physiology and Pathology*.

- Jöns Jacob Berzelius publishes *Traité de chimie minérale, végétale et animal* (6 vols., 1845–50)

- Jean-Baptiste Boussingault publishes *Agronomie, chimie agricole, et physiologie* (5 vols., 1860–1874; 2nd ed., 1884).

- In 1868 Samuel William Johnson publishes *How Crops Grow*.

- In 1870 S. W. Johnson publishes *How Crops Feed: A treatise on the atmosphere and soil as related to the nutrition of agricultural plants*.

- In 1872 Karl Heinrich Ritthausen publishes *Protein bodies in grains, legumes, and linseed. Contributions to the physiology of seeds for cultivation, nutrition, and fodder*

Agricultural Diversification

In the agricultural context, diversification can be regarded as the re-allocation of some of a farm's productive resources, such as land, capital, farm equipment and pices to other farmers and, particularly in richer countries, non-farming activities such as restaurants and shops. Factors leading to decisions to diversify are many, but include; reducing risk, responding to changing consumer demands or changing government policy, responding to external shocks and, more recently, as a consequence of climate change.

Definitions of Diversification

While most definitions of diversification in developing countries do work on the assumption that diversification primarily involves a substitution of one crop or other agricultural product for another, or an increase in the number of enterprises, or activities, carried out by a particular farm, the definition used in developed countries sometimes relates more to the development of activities on the farm that do not involve agricultural production. For example, one section of the British Department for Environment, Food and Rural Affairs (DEFRA) defines diversification as "the entrepreneurial use of farm resources for a non-agricultural purpose for commercial gain". Using this definition DEFRA found that 56% of UK farms had diversified in 2003. The great majority of diversification activities simply involved the renting out of farm buildings for non-farming use, but 9% of farms had become involved with processing or retailing, 3% with provision of tourist accommodation or catering, and 7% with sport or recreational activities. Others adopt a broader definition, which may include development of new marketing opportunities.

In developing countries such as India, which has been one of the leaders in promoting diversification, the concept is applied both to individual farmers and to different regions, with government programmes being aimed at promoting widespread diversification. The concept in India is seen as referring to the "shift from the regional dominance of one crop to regional production of a number of crops ...... (which takes into account)..... the economic returns from different value-added crops... with complementary marketing opportunities".

Drivers of Diversification

Diversification can be a response to both opportunities and threats.

Opportunities

- Changing consumer demand. As consumers in developing countries become richer, food consumption patterns change noticeably. People move away from a diet based on staples to one with a greater content of animal products (meat, eggs, and dairy) and fruits and vegetables. In turn, more dynamic farmers are able to diversify to meet these needs.

- Changing demographics. Rapid urbanization in developing countries affects consumption patterns. Moreover, a smaller number of farmers, in percentage terms at least, has to supply a larger number of consumers. While this may not imply diversification it does require adaptation to new farming techniques to meet higher demand.

- Export potential. Developing country farmers have had considerable success by diversifying into crops that can meet export market demand. While concern about food miles, as well as the cost of complying with supermarket certification requirements such as for GlobalGAP may jeopardize this success in the long run, there remains much potential to diversify to meet export markets.

- Adding value. The pattern witnessed in the West, and now becoming widespread in developing countries, is for consumers to devote less and less time to food preparation. They increasingly require ready-prepared meals and labour-saving packaging, such as pre-cut salads. This provides the opportunity for farmers to diversify into value addition, particularly in countries where supermarkets play a major role in retailing.

- Changing marketing opportunities. The changing of government policies that control the way in which farmers can link to markets can open up new diversification possibilities. For example, in India, policy changes to remove the monopoly of state "regulated markets" to handle all transactions made it possible for farmers to establish direct contracts with buyers for new products.

- Improving nutrition. Diversifying from the monoculture of traditional staples can have important nutritional benefits for farmers in developing countries.

Threats

- Urbanization. This is both an opportunity and a threat, in that the expansion of cities places pressure on land resources and puts up the value of the land. If farmers are to remain on the land they need to generate greater income from that land than they could by growing basic staples. This fact, and the proximity of markets, explains why farmers close to urban areas tend to diversify into high-value crops.

- Risk. Farmers face risk from bad weather and from fluctuating prices. Diversification is a logical response to both. For example, some crops are more drought-resistant than others, but may offer poorer economic returns. A diversified portfolio of products should ensure that farmers do not suffer complete ruin when the weather is bad. Similarly, diversification can manage price risk, on the assumption that not all products will suffer low prices at the same time. In fact, farmers often do the opposite of diversification by planting products that have a high price in one year, only to see the price collapse in the next, as explained by the cobweb theory.

- External threats. Farmers who are dependent on exports run the risk that conditions will change in their markets, not because of a change in consumer demand but because of policy changes. A classic example is the Caribbean banana industry, which collapsed as a result of the removal of quota protection on EU markets, necessitating diversification by the region's farmers.

- Domestic policy threats. Agricultural production is sometimes undertaken as a consequence of government subsidies, rather than because it is inherently profitable. The reduction or removal of those subsidies, whether direct or indirect, can have a major effect on farmers and provide a significant incentive for diversification or, in some cases, for returning to production of crops grown prior to the introduction of subsidies.

- Climate change. The type of crop that can be grown is affected by changes in temperatures and the length of the growing season. Climate change could also modify the availability of water for production. Farmers in several countries, including Canada, India, Kenya, Mozambique, and Sri Lanka have already initiated diversification as a response to climate change. Government policy in Kenya to promote crop diversification has included the removal of subsidies for some crops, encouraging land-use zoning and introducing differential land tax systems.

Opportunities for Diversification

In making decisions about diversification farmers need to consider whether income generated by new farm enterprises will be greater than the existing activities, with similar or less risk. While growing new crops or raising animals may be technically possible, these may not be suitable for many farmers in terms of their land, labour and capital resources. Moreover, markets for the products may be lacking. The United Nations Food and Agriculture Organization (FAO) has been one of the development organizations promoting diversification by small farmers and has produced booklets identifying beekeeping, mushroom farming, milk production, fish ponds and sheep and goats, among others, as diversification possibilities.

Measures of Diversification

Agricultural diversification is measured in a number of ways throughout the world. For example, one such measure is the *index of maximum proportion*, which is "defined as the ratio (proportion) of the farm's primary activity to its total activities".

Agricultural Economics

Agricultural economics or agronomics is an applied field of economics concerned with the application of economic theory in optimizing the production and distribution of food and fibre—a discipline known as agronomics. Agronomics was a branch of economics that specifically dealt with land usage. It focused on maximizing the crop yield while maintaining a good soil ecosystem. Throughout the 20th century the discipline expanded and the current scope of the discipline is much broader. Agricultural economics today includes a variety of applied areas, having considerable overlap with conventional economics. Agricultural economists have made substantial contributions to research in economics, econometrics, development economics, and environmental economics. Agricultural economics influences food policy, agricultural policy, and environmental policy.

Origins

Economics has been defined as the study of resource allocation under scarcity. Agronomics, or the application of economic methods to optimizing the decisions made by agricultural producers, grew to prominence around the turn of the 20th century. The field of agricultural economics can be traced out to works on land economics. Henry Charles Taylor was the greatest contributor with the establishment of the Department of Agricultural Economics at Wisconsin in 1909.

Another contributor, 1979 Nobel Economics Prize winner Theodore Schultz, was among the first to examine development economics as a problem related directly to agriculture. Schultz was also instrumental in establishing econometrics as a tool for use in analyzing agricultural economics empirically; he noted in his landmark 1956 article that agricultural supply analysis is rooted in "shifting sand", implying that it was and is simply not being done correctly.

One scholar summarizes the development of agricultural economics as follows:

"Agricultural economics arose in the late 19th century, combined the theory of the firm with marketing and organization theory, and developed throughout the 20th century largely as an empirical branch of general economics. The discipline was closely linked to empirical applications of mathematical statistics and made early and significant contributions to econometric methods. In the 1960s and afterwards, as agricultural sectors in the OECD countries contracted, agricultural economists were drawn to the development problems of poor countries, to the trade and macroeconomic policy implications of agriculture in rich countries, and to a variety of production, consumption, and environmental and resource problems."

Agricultural economists have made many well-known contributions to the economics field with such models as the cobweb model, hedonic regression pricing models, new technology and diffusion models (Zvi Griliches), multifactor productivity and efficiency theory and measurement, and the random coefficients regression. The farm sector is frequently cited as a prime example of the perfect competition economic paradigm.

In Asia, agricultural economics was offered first by the University of the Philippines Los Baños Department of Agricultural Economics in 1919. Today, the field of agricultural economics has transformed into a more integrative discipline which covers farm management and production economics, rural finance and institutions, agricultural marketing and prices, agricultural policy and development, food and nutrition economics, and environmental and natural resource economics.

Since the 1970s, agricultural economics has primarily focused on seven main topics, according to a scholar in the field: agricultural environment and resources; risk and uncertainty; food and consumer economics; prices and incomes; market structures; trade and development; and technical change and human capital.

Major Topics in Agricultural Economics

Agricultural Environment and Natural Resources

In the field of environmental economics, agricultural economists have contributed in three main areas: designing incentives to control environmental externalities (such as water pollution due to agricultural production), estimating the value of non-market benefits from natural resources and environmental amenities (such as an appealing rural landscape), and the complex interrelationship between economic activities and environmental consequences. With regard to natural resources, agricultural economists have developed quantitative tools for improving land management, preventing erosion, managing pests, protecting biodiversity, and preventing livestock diseases.

Food and Consumer Economics

While at one time, the field of agricultural economics was focused primarily on farm-level issues, in recent years agricultural economists have studied diverse topics related to the economics of food consumption. In addition to economists' long-standing emphasis on the effects of prices and incomes, researchers in this field have studied how information and quality attributes influence consumer behavior. Agricultural economists have contributed to understanding how households make choices between purchasing food or preparing it at home, how food prices are determined, definitions of poverty thresholds, how consumers respond to price and income changes in a consistent way, and survey and experimental tools for understanding consumer preferences.

Production Economics and Farm Management

Agricultural economics research has addressed diminishing returns in agricultural production, as well as farmers' costs and supply responses. Much research has applied economic theory to farm-level decisions. Studies of risk and decision-making under uncertainty have real-world applications to crop insurance policies and to understanding how farmers in developing countries make choices about technology adoption. These topics are important for understanding prospects for producing sufficient food for a growing world population, subject to new resource and environmental challenges such as water scarcity and global climate change.

Development Economics

Development economics is broadly concerned with the improvement of living conditions in low-income countries, and the improvement of economic performance in low-income settings. Because agriculture is a large part of most developing economies, both in terms of employment and share of GDP, agricultural economists have been at the forefront of empirical research on development economics, contributing to our understanding of agriculture's role in economic development, economic growth and structural transformation. Many agricultural economists are interested in the food systems of developing economies, the linkages between agriculture and nutrition, and the ways in which agriculture interact with other domains, such as the natural environment.

Professional Associations

The International Association of Agricultural Economists (IAAE) is a worldwide professional association, which holds its major conference once every three years. The association publishes the journal *Agricultural Economics*. There also is a European Association of Agricultural Economists (EAAE), an African Association of Agricultural Economists [AAAE]and an Australian Agricultural and Resource Economics Society. Substantial work in agricultural economics internationally is conducted by the International Food Policy Research Institute.

In the United States, the primary professional association is the Agricultural & Applied Economics Association (AAEA), which holds its own annual conference and also co-sponsors the annual meetings of the Allied Social Sciences Association (ASSA). The AAEA publishes the American Journal of Agricultural Economics and Applied Economic Perspectives and Policy.

Careers in Agricultural Economics

Graduates from agricultural and applied economics departments find jobs in many sectors of the economy: agricultural management, agribusiness, commodities markets, education, the financial sector, government, natural resource and environmental management, real estate, and public relations. Careers in agricultural economics require at least a bachelor's degree, and research careers in the field require graduate-level training. A 2011 study by the Georgetown Center on Education and the Workforce rated agricultural economics tied for 8th out of 171 fields in terms of employability.

Literature

Evenson, Robert E. and Prabhu Pingali (eds.) (2007). *Handbook of Agricultural Economics*. Amsterdam, NL: Elsevier.

Agricultural Philosophy

Agricultural philosophy (or philosophy of agriculture) is, roughly and approximately, a discipline devoted to the systematic critique of the philosophical frameworks (or ethical world views) that are the foundation for decisions regarding agriculture. Many of these views are also used to guide decisions dealing with land use in general. In everyday usage, it can also be defined as the love of, search after, and wisdom associated with agriculture, as one of humanity's founding components of civilization. However, this view is more aptly known as agrarianism when it refers to a lifestyle preference, often somewhat romantic in orientation that can lead to unfounded beliefs that may be difficult to shiftsee video In actuality, agrarianism is only one philosophy or normative framework out of many that people use to guide their decisions regarding agriculture on an everyday basis. The most prevalent of these philosophies will be briefly defined below.

Utilitarian Approach

This view was first put forth by Jeremy Bentham and John Stuart Mill. Though there are many varieties of utilitarianism, generally the view is that a morally right action is an action that produces the maximum good for people. This theory is a form of consequentialism; which basically means that the correct action is understood entirely in terms of the consequences of that action. Utilitarianism is often used when deciding farming issues. For example, farmland is commonly valued based upon its capacity to the grow crops that people want. This approach to valuing land is called Asset Theory (in contrast to Location Theory) and it is based upon utilitarian principles. Another example is when a community decides on what to do with a particular parcel of land. Let's say that this community must decide to use it for industry, residential uses, or for farming. By using a utilitarian approach, the council would judge which use would benefit the greatest number of people in the community and then make their choice based upon that information. Finally, it also forms the foundation for industrial farming; as an increase in yield, which would increase the number of people able to receive goods from farmed land, is judged from this view to be a good action or approach. Indeed, a common argument in favor of industrial agriculture is this it is a good practice because it increases the benefits for humans; benefits such as food abundance and a drop in food prices.

However, several scholars and writers, such as Peter Singer, Aldo Leopold, Vandana Shiva, Barbara Kingsolver, and Wendell Berry have argued against this view. For example, Singer argues that the suffering of animals (farm animals included) should be included in the cost/benefit calculus when deciding whether or not to do an action such as industrial farming. It has also been challenged on the grounds that farmland and farm animals are instrumentalized in this view and not valued in and of themselves. In addition, systems thinkers, deep ecologists, and agrarian philosophers (such as Aldo Leopold & Wendell Berry) critique this view on the grounds that it ignores aspects of farming which are morally applicable and/or intrinsically valuable. The Slow Food Movement and the Buy Local Agricultural Movements are also built upon philosophical views morally opposed to extreme versions of this approach. Other critiques will be explored below when different philosophical approaches to agriculture are briefly explained. However, it is important to note that the Utilitarian approach to agriculture is currently the most widespread approach within the modern Western World.

Libertarian Approach

Another philosophical approach often used when deciding land or farming issues is Libertarianism. Libertarianism is, roughly, the moral view that agents own themselves and have certain moral rights including the right to acquire property. In a looser sense, Libertarianism is commonly identified with the belief that each person has a right to a maximum amount of liberty when this liberty does not interfere with other people's freedom. A well known Libertarian theorist is John Hospers. Within this view, property rights are natural rights. Thus, it would be acceptable for a farmer to inefficiently farm their land as long as they don't harm others while doing it. In 1968, Garrett Harden applied this philosophy to land/farming issues when he argued that the only solution to the "Tragedy of the Commons" was to place soil and water resources into the hands of private citizens. He then supplied utilitarian justifications to support his argument and, indeed, you could argue that Libertarianism is rooted in utilitarian ideals. However, this leaves Libertarian based land ethics open to the above critiques lodged against Utilitarian approaches to agriculture. Even excepting these critiques, the Libertarian view has been specifically challenged by the critique that people making self-interested decisions can cause large ecological and social disasters such as the Dust Bowl disaster. Even so, it is a philosophical view commonly held within the United States and, especially, by U.S. ranchers and farmers.

Egalitarian Approach

Egalitarian based views are often developed as a response to Libertarianism. This is because, while Libertarianism provides for the maximum amount of human freedom, it does not require a person to help others. In addition, it also leads to the grossly uneven distribution of wealth. A well known Egalitarian philosopher is John Rawls. When focusing on agriculture, what this translates into is the uneven distribution of land and food. While both Utilitarian and Libertarian approaches to agriculture ethics could conceivably rationalize this mal-distribution, an Egalitarian approach typically favors equality whether that be equal entitlement and/or opportunity to employment or access to food. However, if you recognize that people have a right to something, then someone has to supply this opportunity or item; whether that be an individual person or the government. Thus, the Egalitarian view links land and water with the right to food. With the growth of human populations and the decline of soil and water resources, Egalitarianism could provide a strong argument for the preservation of soil fertility and water.

Ecological or Systems Approach

In addition to Utilitarian, Libertarian, and Egalitarian philosophies, there are also normative views that are based upon the principle that land has intrinsic value and positions coming out of an ecological or systems view. Two main examples of this are James Lovelock's Gaia hypothesis which postulates that the Earth is an organism and deep ecologists who argue that human communities are built upon a foundation of the surrounding ecosystems or the biotic communities. While the above philosophies can be useful for guiding decision making on issues concerning land in general, they have limited usefulness when applied to agriculture because these philosophies privilege natural ecosystems and agricultural ecosystems are often considered not natural. One philosophy grounded in the principle that land has intrinsic value which is directly applicable to agriculture is Aldo Leopold's stewardship ethic or land ethic. For Leopold, an action is correct if it tends to "preserve the integrity, stability, and beauty of the biotic community". Similar to Egalitarian-based land ethics, many of the above philosophies were also developed as alternatives to utilitarian and libertarian based approaches. Leopold's ethic is currently one of the most popular ecological approaches to agriculture commonly known as agrarianism. Other agrarianists include Benjamin Franklin, Thomas Jefferson, J. Hector St. John de Crèvecœur (1735–1813), Ralph Waldo Emerson (1803–1882), Henry David Thoreau (1817–1862), John Steinbeck (1902–1968), Wendell Berry (b. 1934), Gene Logsdon (b. 1932), Paul B. Thompson, and Barbara Kingsolver. One contemporary approach embodying these elements in agricultural aid for developing countries is international integrated development, which builds on the United Nations' SDGs.

Agricultural Marketing

Market display in China

Agricultural marketing covers the services involved in moving an agricultural product from the farm to the consumer. Numerous interconnected activities are involved in doing this, such as planning production, growing and harvesting, grading, packing, transport, storage, agro- and food processing, distribution, advertising and sale. Some definitions would even include "the acts of buying supplies, renting equipment, (and) paying labor", arguing that marketing is everything a business does. Such activities cannot take place without the exchange of information and are often heavily dependent on the availability of suitable finance.

Marketing systems are dynamic; they are competitive and involve continuous change and improvement. Businesses that have lower costs, are more efficient, and can deliver quality products, are those that prosper. Those that have high costs, fail to adapt to changes in market demand and provide poorer quality are often forced out of business. Marketing has to be customer-oriented and has to provide the farmer, transporter, trader, processor, etc. with a profit. This requires those involved in marketing chains to understand buyer requirements, both in terms of product and business conditions.

In Western countries considerable agricultural marketing support to farmers is often provided. In the USA, for example, the USDA operates the Agricultural Marketing Service. Support to developing countries with agricultural marketing development is carried out by various donor organizations and there is a trend for countries to develop their own Agricultural Marketing or Agribusiness units, often attached to ministries of agriculture. Activities include market information development, marketing extension, training in marketing and infrastructure development. Since the 1990s trends have seen the growing importance of supermarkets and a growing interest in contract farming, both of which impact significantly on the way in which marketing takes place.

Agricultural Marketing Support

In the United States the Agricultural Marketing Service (AMS) is a division of USDA and has programs for cotton, dairy, fruit and vegetable, livestock and seed, poultry, and tobacco. These programs provide testing, standardization, grading and market news services and oversee marketing agreements and orders, administer research and promotion programs, and purchase commodities for federal food programs. The AMS also enforces certain federal laws. USDA also provides support to the Agricultural Marketing Resource Center at Iowa State University and to Penn State University.

In the United Kingdom support for marketing of some commodities was provided before and after the Second World War by boards such as the Milk Marketing Board and the Egg Marketing Board, but these were closed down in the 1970s. As a colonial power Britain established marketing boards in many countries, particularly in Africa. Some continue to exist although many were closed down at the time of the introduction of structural adjustment measures in the 1990s.

In recent years several developing countries have established government-sponsored marketing or agribusiness units. South Africa, for example, started the National Agricultural Marketing Council (NAMC) as a response to the deregulation of the agriculture industry and closure of marketing boards in the country. India has the long-established National Institute of Agricultural Marketing (NIAM). These are primarily research and policy organizations, but other agencies provide facilitating services for marketing channels, such as the provision of infrastructure, market information and documentation support. Examples include the National Agricultural Marketing Development Corporation (NAMDEVCO) in Trinidad and Tobago and the New Guyana Marketing Corporation.

Several organizations provide support to developing countries to develop their agricultural marketing systems, including FAO's agricultural marketing unit and various donor organizations. There has also recently been considerable interest by NGOs to carry out activities to link farmers to markets.

Agricultural Marketing Development

Congestion at a market in Abidjan

Well-functioning marketing systems necessitate a strong private sector backed up by appropriate policy and legislative frameworks and effective government support services. Such services can include provision of market infrastructure, supply of market information (as done by USDA, for example), and agricultural extension services able to advise farmers on marketing. Training in marketing at all levels is also needed. One of many problems faced in agricultural marketing in developing countries is the latent hostility to the private sector and the lack of understanding of the role of the intermediary. For this reason "middleman" has become very much a pejorative word.

A typical market in Africa

Agricultural Advisory Services and the Market

Promoting market orientation in agricultural advisory services aims to provide for the sustainable enhancement of the capabilities of the rural poor to enable them to benefit from agricultural markets and help them to adapt to factors which impact upon these. As a study by the Overseas Development Institute demonstrates, a value chain approach to advisory services indicates that the range of clients serviced should go beyond farmers to include input providers, producers, producer organizations and processors and traders.

Market Infrastructure

Efficient marketing infrastructure such as wholesale, retail and assembly markets and storage facilities is essential for cost-effective marketing, to minimize post-harvest losses and to reduce health risks. Markets play an important role in rural development, income generation, food security, developing rural-market linkages and gender issues. Planners need to be aware of how to design markets that meet a community's social and economic needs and how to choose a suitable site for a new market. In many cases sites are chosen that are inappropriate and result in under-use or even no use of the infrastructure constructed. It is also not sufficient just to build a market: attention needs to be paid to how that market will be managed, operated and maintained. In most cases, where market improvements were only aimed at infrastructure upgrading and did not guarantee maintenance and management, most failed within a few years.

Rural assembly markets are located in production areas and primarily serve as places where farmers can meet with traders to sell their products. These may be occasional (perhaps weekly) markets, such as haat bazaars in India and Nepal, or permanent. Terminal wholesale markets are located in major metropolitan areas, where produce is finally channelled to consumers through trade between wholesalers and retailers, caterers, etc. The characteristics of wholesale markets have changed considerably as retailing changes in response to urban growth, the increasing role of supermarkets and increased consumer spending capacity. These changes require responses in the way in which traditional wholesale markets are organized and managed.

Retail marketing systems in western countries have broadly evolved from traditional street markets through to the modern hypermarket or out-of-town shopping center. In developing countries, there remains considerable scope to improve agricultural marketing by constructing new retail markets, despite the growth of supermarkets, although municipalities often view markets as sources of revenue rather than infrastructure requiring development. Effective regulation of markets is essential. Inside the market, both hygiene rules and revenue collection activities have to be enforced. Of equal importance, however, is the maintenance of order outside the market. Licensed traders in a market will not be willing to cooperate in raising standards if they face competition from unlicensed operators outside who do not pay any of the costs involved in providing a proper service.

Market Information

Efficient market information can be shown to have positive benefits for farmers and traders. Up-to-date information on prices and other market factors enables farmers to negotiate with traders and also facilitates spatial distribution of products from rural areas to towns and between markets. Most governments in developing countries have tried to provide market information services to farmers, but these have tended to experience problems of sustainability. Moreover, even when they function, the service provided is often insufficient to allow commercial decisions to be made because of time lags between data collection and dissemination. Modern communications technologies open up the possibility for market information services to improve information delivery through SMS on cell phones and the rapid growth of FM radio stations in many developing countries offers the possibility of more localised information services. In the longer run, the internet may become an effective way of delivering information to farmers. However, problems associated with the cost and accuracy of data collection still remain to be addressed. Even when they have access to market information, farmers often require assistance in interpreting that information. For example, the market price quoted on the radio may refer to a wholesale selling price and farmers may have difficulty in translating this into a realistic price at their local assembly market. Various attempts have been made in developing countries to introduce commercial market information services but these have largely been targeted at traders, commercial farmers or exporters. It is not easy to see how small, poor farmers can generate sufficient income for a commercial service to be profitable although in India a new service introduced by Thomson Reuters was reportedly used by over 100,000 farmers in its first year of operation. Esoko in West Africa attempts to subsidize the cost of such services to farmers by charging access to a more advanced feature set of mobile-based tools to businesses.

Marketing Training

Farmers frequently consider marketing as being their major problem. However, while they are able to identify such problems as poor prices, lack of transport and high post-harvest losses, they are often poorly equipped to identify potential solutions. Successful marketing requires learning new skills, new techniques and new ways of obtaining information. Extension officers working with ministries of agriculture or NGOs are often well-trained in horticultural production techniques but usually lack knowledge of marketing or post-harvest handling. Ways of helping them develop their knowledge of these areas, in order to be better able to advise farmers about market-oriented horticulture, need to be explored. While there is a range of generic guides and other training materials available from FAO and others, these should ideally be tailored to national circumstances to have maximum effect.

Enabling Environments

Agricultural marketing needs to be conducted within a supportive policy, legal, institutional, macro-economic, infrastructural and bureaucratic environment. Traders and others cannot make investments in a climate of arbitrary government policy changes, such as those that restrict imports and exports or internal produce movement. Those in business cannot function if their trading activities are hampered by excessive bureaucracy. Inappropriate law can distort and reduce the efficiency of the market, increase the costs of doing business and retard the development of a competitive private sector. Poor support institutions, such as agricultural extension services, municipalities that operate markets inefficiently and export promotion bodies, can be particularly damaging. Poor roads increase the cost of doing business, reduce payments to farmers and increase prices to consumers. Finally, the ever-present problem of corruption can seriously impact on agricultural marketing efficiency in many countries by increasing the transaction costs faced by those in the marketing chain.

Recent Developments

New marketing linkages between agribusiness, large retailers and farmers are gradually being developed, e.g. through contract farming, group marketing and other forms of collective action. Donors and NGOs are paying increasing attention to ways of promoting direct linkages between farmers and buyers within a value chain context. More attention is now being paid to the development of regional markets (e.g. East Africa) and to structured trading systems that should facilitate such developments. The growth of supermarkets, particularly in Latin America and East and South East Asia, is having a significant impact on marketing channels for horticultural, dairy and livestock products. Nevertheless, "spot" markets will continue to be important for many years, necessitating attention to infrastructure improvement such as for retail and wholesale markets.

Agroecology

A community-supported agriculture share of crops

Agroecology is the study of ecological processes applied to agricultural production systems. The prefix *agro-* refers to *agriculture*. Bringing ecological principles to bear in agroecosystems can suggest novel management approaches that would not otherwise be considered. The term is often used imprecisely and may refer to "a science, a movement, [or] a practice". Agroecologists study a variety of agroecosystems, and the field of agroecology is not associated with any one particular method of farming, whether it be organic, integrated, or conventional; intensive or extensive. Although it has much more common thinking and principles with some of the before mentioned farming systems.

Ecological Strategy

Agroecologists do not unanimously oppose technology or inputs in agriculture but instead assess how, when, and if technology can be used in conjunction with natural, social and human assets. Agroecology proposes a context- or site-specific manner of studying agroecosystems, and as such, it recognizes that there is no universal formula or recipe for the success and maximum well-being of an agroecosystem. Thus, agroecology is not defined by certain management practices, such as the use of natural enemies in place of insecticides, or polyculture in place of monoculture.

Instead, agroecologists may study questions related to the four system properties of agroecosystems: productivity, stability, sustainability and equitability. As opposed to disciplines that are concerned with only one or some of these properties, agroecologists see all four properties as interconnected and integral to the success of an agroecosystem. Recognizing that these properties are found on varying spatial scales, agroecologists do not limit themselves to the study of agroecosystems at any one scale: gene-organism-population-community-ecosystem-landscape-biome, field-farm-community-region-state-country-continent-global.

Agroecologists study these four properties through an interdisciplinary lens, using natural sciences to understand elements of agroecosystems such as soil properties and plant-insect interactions, as well as using social sciences to understand the effects of farming practices on rural communities, economic constraints to developing new production methods, or cultural factors determining farming practices.

Approaches

Agroecologists do not always agree about what agroecology is or should be in the long-term. Different definitions of the term agroecology can be distinguished largely by the specificity with which one defines the term "ecology", as well as the term's potential political connotations. Definitions of agroecology, therefore, may be first grouped according to the specific contexts within which they situate agriculture. Agroecology is defined by the OECD as "the study of the relation of agricultural crops and environment." This definition refers to the "-ecology" part of "agroecology" narrowly as the natural environment. Following this definition, an agroecologist would study agriculture's various relationships with soil health, water quality, air quality, meso- and micro-fauna, surrounding flora, environmental toxins, and other environmental contexts.

A more common definition of the word can be taken from Dalgaard et al., who refer to agroecology as the study of the interactions between plants, animals, humans and the environment within agricultural systems. Consequently, agroecology is inherently multidisciplinary, including factors from agronomy, ecology, sociology, economics and related disciplines. In this case, the "-ecology" portion of "agroecology is defined broadly to include social, cultural, and economic contexts as well. Francis et al. also expand the definition in the same way, but put more emphasis on the notion of food systems.

Agroecology is also defined differently according to geographic location. In the global south, the term often carries overtly political connotations. Such political definitions of the term usually ascribe to it the goals of social and economic justice; special attention, in this case, is often paid to the traditional farming knowledge of indigenous populations. North American and European uses of the term sometimes avoid the inclusion of such overtly political goals. In these cases, agroecology is seen more strictly as a scientific discipline with less specific social goals.

Agro-population Ecology

This approach is derived from the science of ecology primarily based on population ecology, which over the past three decades has been displacing the ecosystems biology of Odum. Buttel explains the main difference between the two categories, saying that "the application of population ecology to agroecology involves the primacy not only of analyzing agroecosystems from the perspective of the population dynamics of their constituent species, and their relationships to climate and biogeochemistry, but also there is a major emphasis placed on the role of genetics."

Inclusive Agroecology

Rather than viewing agroecology as a subset of agriculture, Wojtkowski takes a more encompassing perspective. In this, natural ecology and agroecology are the major headings under ecology. Natural ecology is the study of organisms as they interact with and within natural environments. Correspondingly, agroecology is the basis for the land-use sciences. Here humans are the primary governing force for organisms within planned and managed, mostly terrestrial, environments.

As key headings, natural ecology and agroecology provide the theoretical base for their respective sciences. These theoretical bases overlap but differ in a major way. Economics has no role in the functioning of natural ecosystems whereas economics sets direction and purpose in agroecology.

Under agroecology are the three land-use sciences, agriculture, forestry, and agroforestry. Although these use their plant components in different ways, they share the same theoretical core.

Beyond this, the land-use sciences further subdivide. The subheadings include agronomy, organic farming, traditional agriculture, permaculture, and silviculture. Within this system of subdivisions, agroecology is philosophically neutral. The importance lies in providing a theoretical base hitherto lacking in the land-use sciences. This allows progress in biocomplex agroecosystems including the multi-species plantations of forestry and agroforestry.

Applications

To arrive at a point of view about a particular way of farming, an agroecologist would first seek to understand the contexts in which the farm(s) is(are) involved. Each farm may be inserted in a unique combination of factors or contexts. Each farmer may have their own premises about the meanings of an agricultural endeavor, and these meanings might be different from those of agroecologists. Generally, farmers seek a configuration that is viable in multiple contexts, such as family, financial, technical, political, logistical, market, environmental, spiritual. Agroecologists want to understand the behavior of those who seek livelihoods from plant and animal increase, acknowledging the organization and planning that is required to run a farm.

Views on Organic and Non-organic Milk Production

Because organic agriculture proclaims to sustain the health of soils, ecosystems, and people, it has much in common with Agroecology; this does not mean that Agroecology is synonymous with organic agriculture, nor that Agroecology views organic farming as the 'right' way of farming. Also, it is important to point out that there are large differences in organic standards among countries and certifying agencies.

Three of the main areas that agroecologists would look at in farms, would be: the environmental impacts, animal welfare issues, and the social aspects.

Environmental impacts caused by organic and non-organic milk production can vary significantly. For both cases, there are positive and negative environmental consequences.

Compared to conventional milk production, organic milk production tends to have lower eutrophication potential per ton of milk or per hectare of farmland, because it potentially reduces leaching of nitrates (NO_3^-) and phosphates (PO_4^-) due to lower fertilizer application rates. Because organic milk production reduces pesticides utilization, it increases land use per ton of milk due to decreased crop yields per hectare. Mainly due to the lower level of concentrates given to cows in organic herds, organic dairy farms generally produce less milk per cow than conventional dairy farms. Because of the increased use of roughage and the, on-average, lower milk production level per cow, some research has connected organic milk production with increases in the emission of methane.

Animal welfare issues vary among dairy farms and are not necessarily related to the way of producing milk (organically or conventionally).

A key component of animal welfare is freedom to perform their innate (natural) behavior, and this is stated in one of the basic principles of organic agriculture. Also, there are other aspects of animal welfare to be considered – such as freedom from hunger, thirst, discomfort, injury, fear, distress, disease and pain. Because organic standards require loose housing systems, adequate bedding, restrictions on the area of slatted floors, a minimum forage proportion in the ruminant diets, and tend to limit stocking densities both on pasture and in housing for dairy cows, they potentially promote good foot and hoof health. Some studies show lower incidence of placenta retention, milk fever, abomasums displacement and other diseases in organic than in conventional dairy herds. However, the level of infections by parasites in organically managed herds is generally higher than in conventional herds.

Social aspects of dairy enterprises include life quality of farmers, of farm labor, of rural and urban communities, and also includes public health.

Both organic and non-organic farms can have good and bad implications for the life quality of all the different people involved in that food chain. Issues like labor conditions, labor hours and labor rights, for instance, do not depend on the organic/non-organic characteristic of the farm; they can be more related to the socio-economical and cultural situations in which the farm is inserted, instead.

As for the public health or food safety concern, organic foods are intended to be healthy, free of contaminations and free from agents that could cause human diseases. Organic milk is meant to have no chemical residues to consumers, and the restrictions on the use of antibiotics and chemicals in organic food production has the purpose to accomplish this goal. Although dairy cows in both organic and conventional farming practices can be exposed to pathogens, it has been shown that, because antibiotics are not permitted as a preventative measure in organic practices, there are far fewer antibiotic resistant pathogens on organic farms. This dramatically increases the efficacy of antibiotics when/if they are necessary.

In an organic dairy farm, an agroecologist could evaluate the following:

1. Can the farm minimize environmental impacts and increase its level of sustainability, for instance by efficiently increasing the productivity of the animals to minimize waste of feed and of land use?

2. Are there ways to improve the health status of the herd (in the case of organics, by using biological controls, for instance)?

3. Does this way of farming sustain good quality of life for the farmers, their families, rural labor and communities involved?

Views on No-till Farming

No-tillage is one of the components of conservation agriculture practices and is considered more environmental friendly than complete tillage. There is a general consensus that no-till can increase soils capacity of acting as a carbon sink, especially when combined with cover crops.

No-till can contribute to higher soil organic matter and organic carbon content in soils, though reports of no-effects of no-tillage in organic matter and organic carbon soil contents also exist, depending on environmental and crop conditions. In addition, no-till can indirectly reduce CO_2 emissions by decreasing the use of fossil fuels.

Most crops can benefit from the practice of no-till, but not all crops are suitable for complete no-till agriculture. Crops that do not perform well when competing with other plants that grow in untilled soil in their early stages can be best grown by using other conservation tillage practices, like a combination of strip-till with no-till areas. Also, crops which harvestable portion grows underground can have better results with strip-tillage, mainly in soils which are hard for plant roots to penetrate into deeper layers to access water and nutrients.

The benefits provided by no-tillage to predators may lead to larger predator populations, which is a good way to control pests (biological control), but also can facilitate predation of the crop itself. In corn crops, for instance, predation by caterpillars can be higher in no-till than in conventional tillage fields.

In places with rigorous winter, untilled soil can take longer to warm and dry in spring, which may delay planting to less ideal dates. Another factor to be considered is that organic residue from the prior year's crops lying on the surface of untilled fields can provide a favorable environment to pathogens, helping to increase the risk of transmitting diseases to the future crop. And because no-till farming provides good environment for pathogens, insects and weeds, it can lead farmers to a more intensive use of chemicals for pest control. Other disadvantages of no-till include underground rot, low soil temperatures and high moisture.

Based on the balance of these factors, and because each farm has different problems, agroecologists will not atest that only no-till or complete tillage is the right way of farming. Yet, these are not the only possible choices regarding soil preparation, since there are intermediate practices such as strip-till, mulch-till and ridge-till, all of them – just as no-till – categorized as conservation tillage. Agroecologists, then, will evaluate the need of different practices for the contexts in which each farm is inserted.

In a no-till system, an agroecologist could ask the following:

1. Can the farm minimize environmental impacts and increase its level of sustainability; for instance by efficiently increasing the productivity of the crops to minimize land use?

2. Does this way of farming sustain good quality of life for the farmers, their families, rural labor and rural communities involved?

History

Pre-WWII

The notions and ideas relating to crop ecology have been around since at least 1911 when F.H. King released *Farmers of Forty Centuries*. King was one of the pioneers as a proponent of more quantitative methods for characterization of water relations and physical properties of soils. In the late 1920s the attempt to merge agronomy and ecology was born with the development of the field of crop ecology. Crop ecology's main concern was where crops would be best grown. Actually, it was only in 1928 that agronomy and ecology were formally linked by Klages.

The first mention of the term agroecology was in 1928, with the publication of the term by Bensin in 1928. The book of Tischler (1965), was probably the first to be actually titled 'agroecology'. He analysed the different components (plants, animals, soils and climate) and their interactions within an agroecosystem as well as the impact of human agricultural management on these components. Other books dealing with agroecology, but without using the term explicitly were published by the German zoologist Friederichs (1930) with his book on agricultural zoology and related ecological/environmental factors for plant protection, and by American crop physiologist Hansen in 1939 when both used the word as a synonym for the application of ecology within agriculture.

Post-WWII

Gliessman mentions that post-WWII, groups of scientists with ecologists gave more focus to experiments in the natural environment, while agronomists dedicated their attention to the cultivated systems in agriculture. According to Gliessman, the two groups kept their research and interest apart until books and articles using the concept of agroecosystems and the word agroecology started to appear in 1970. Dalgaard explains the different points of view in ecology schools, and the fundamental differences, which set the basis for the development of agroecology. The early ecology school of Henry Gleason investigated plant populations focusing in the hierarchical levels of the organism under study.

Friederich Clement's ecology school, however included the organism in question as well as the higher hierarchical levels in its investigations, a "landscape perspective". However, the ecological schools where the roots of agroecology lie are even broader in nature. The ecology school of Tansley, whose view included both the biotic organism and their environment, is the one from which the concept of agroecosystems emerged in 1974 with Harper.

In the 1960s and 1970s the increasing awareness of how humans manage the landscape and its consequences set the stage for the necessary cross between agronomy and ecology. Even though, in many ways the environmental movement in the US was a product of the times, the Green Decade,spread an environmental awareness of the unintended consequences of changing ecological processes. Works such as *Silent Spring*, and *The Limits to Growth*, and changes in legislation such as the Clean Air Act, Clean Water Act, and the National Environmental Policy Act caused the public to be aware of societal growth patterns, agricultural production, and the overall capacity of the system.

Fusion With Ecology

After the 1970s, when agronomists saw the value of ecology and ecologists began to use the agricultural systems as study plots, studies in agroecology grew more rapidly. Gliessman describes that the innovative work of Prof. Efraim Hernandez X., who developed research based on indigenous systems of knowledge in Mexico, led to education programs in agroecology. In 1977 Prof. Efraim Hernandez X. explained that modern agricultural systems had lost their ecological foundation when socio-economic factors became the only driving force in the food system. The acknowledgement that the socio-economic interactions are indeed one of the fundamental components of any agroecosystems came to light in 1982, with the article Agroecologia del Tropico Americano by Montaldo. The author argues that the socio-economic context cannot be separated from the agricultural systems when designing agricultural practices.

In 1995 Edens et al. in Sustainable Agriculture and Integrated Farming Systems solidified this idea proving his point by devoting special sections to economics of the systems, ecological impacts, and ethics and values in agriculture. Actually, 1985 ended up being a fertile and creative year for the new discipline. For instance in the same year, Miguel Altieri integrated how consolidation of the farms, and cropping systems impact pest populations. In addition, Gliessman highlighted that socio-economic, technological, and ecological components give rise to producer choices of food production systems. These pioneering agroecologists have helped to frame the foundation of what we today consider the interdisciplinary field of agroecology and have led to advances in a number of farming systems. In Asian rice, for example, crop diversification by growing flowering crops in strips beside rice fields has recently been demonstrated to reduce pests so effectively (by the flower nectar attracting and supporting parasitoids and predators) that insecticide spraying is reduced by 70%, yields increase by 5%, together resulting in an economic advantage of 7.5% (Gurr et al., 2016).

By Region

The principles of agroecology are expressed differently depending on local ecological and social contexts.

Latin America

Latin America's experiences with North American Green Revolution agricultural techniques have opened space for agroecologists. Traditional or indigenous knowledge represents a wealth of possibility for agroecologists, including "exchange of wisdoms". See Miguel Alteiri's *Enhancing the Productivity of Latin American Traditional Peasant Farming Systems Through an Agroecological Approach* for information on agroecology in Latin America.

Madagascar

Most of the historical farming in Madagascar has been conducted by indigenous peoples. The French colonial period disturbed a very small percentage of land area, and even included some useful experiments in Sustainable forestry. Slash-and-burn techniques, a component of some shifting cultivation systems have been practised by natives in Madagascar for centuries. As of 2006 some of the major agricultural products from slash-and-burn methods are wood, charcoal and grass for Zebu grazing. These practices have taken perhaps the greatest toll on land fertility since the end of French rule, mainly due to overpopulation pressures.

Agricultural Soil Science

Agricultural soil science is a branch of soil science that deals with the study of edaphic conditions as they relate to the production of food and fiber. In this context, it is also a constituent of the field of agronomy and is thus also described as soil agronomy.

History

Prior to the development of pedology in the 19th century, agricultural soil science (or edaphology) was the only branch of soil science. The bias of early soil science toward viewing soils only in terms of their agricultural potential continues to define the soil science profession in both academic and popular settings as of 2006. (Baveye, 2006)

Current Status

Agricultural soil science follows the holistic method. Soil is investigated in relation to and as integral part of terrestrial ecosystems but is also recognized as a manageable natural resource.

Agricultural soil science studies the chemical, physical, biological, and mineralogical composition of soils as they relate to agriculture. Agricultural soil scientists develop methods that will improve the use of soil and increase the production of food and fiber crops. Emphasis continues to grow on the importance of soil sustainability. Soil degradation such as erosion, compaction, lowered fertility, and contamination continue to be serious concerns. They conduct research in irrigation and drainage, tillage, soil classification, plant nutrition, soil fertility, and other areas.

Although maximizing plant (and thus animal) production is a valid goal, sometimes it may come at high cost which can be readily evident (e.g. massive crop disease stemming from monoculture) or long-term (e.g. impact of chemical fertilizers and pesticides on human health). An agricultural soil scientist may come up with a plan that can maximize production using sustainable methods and solutions, and in order to do that he must look into a number of science fields including agricultural science, physics, chemistry, biology, meteorology and geology.

Soil Variables

Some soil variables of special interest to agricultural soil science are:

- Soil texture or soil composition: Soils are composed of solid particles of various sizes. In decreasing order, these particles are sand, silt and clay. Every soil can be classified according to the relative percentage of sand, silt and clay it contains.

- Aeration and porosity: Atmospheric air contains elements such as oxygen, nitrogen, carbon and others. These elements are prerequisites for life on Earth. Particularly, all cells (including root cells) require oxygen to function and if conditions become anaerobic they fail to respire and metabolize. Aeration in this context refers to the mechanisms by which air is delivered to the soil. In natural ecosystems soil aeration is chiefly accomplished through the vibrant activity of the biota. Humans commonly aerate the soil by tilling and plowing, yet such practice may cause degradation. Porosity refers to the air-holding capacity of the soil.

- Drainage: In soils of bad drainage the water delivered through rain or irrigation may pool and stagnate. As a result, prevail anaerobic conditions and plant roots suffocate. Stagnant water also favors plant-attacking water molds. In soils of excess drainage, on the other hand, plants don't get to absorb adequate water and nutrients are washed from the porous medium to end up in groundwater reserves.

- Water content: Without soil moisture there is no transpiration, no growth and plants wilt. Technically, plant cells loose their pressure. Plants contribute directly to soil moisture. For instance, they create a leafy cover that minimizes the evaporative effects of solar radiation. But even when plants or parts of plants die, the decaying plant matter produces a thick organic cover that protects the soil from evaporation, erosion and compaction.

- Water potential: Water potential describes the tendency of the water to flow from one area of the soil to another. While water delivered to the soil surface normally flows downward due to gravity, at some point it meets increased pressure which causes a reverse upward flow. This effect is known as water suction.

- Horizonation: Typically found in advanced and mature soils, horizonation refers to the creation of soil layers with differing characteristics. It affects almost all soil variables.

- Fertility: A fertile soil is one rich in nutrients and organic matter. Modern agricultural methods have rendered much of the arable land infertile. In such cases, soil can no longer support on its own plants with high nutritional demand and thus needs an external source of nutrients. However, there are cases where human activity is thought to be responsible for transforming rather normal soils into super-fertile ones.

- Biota and soil biota: Organisms interact with the soil and contribute to its quality in innumerable ways. Sometimes the nature of interaction may be unclear, yet a rule is becoming evident: The amount and diversity of the biota is "proportional" to the quality of the soil. Clades of interest include bacteria, fungi, nematodes, annelids and arthropods.

- Soil acidity or soil pH and cation-exchange capacity: Root cells act as hydrogen pumps and the surrounding concentration of hydrogen ions affects their ability to absorb nutrients. pH is a measure of this concentration. Each plant species achieves maximum growth in a particular pH range, yet the vast majority of edible plants can grow in soil pH between 5.0 and 7.5.

Soil Fertility

Agricultural soil scientists study ways to make soils more productive. They classify soils and test them to determine whether they contain nutrients vital to plant growth. Such nutritional substances include compounds of nitrogen, phosphorus, and potassium. If a certain soil is deficient in these substances, fertilizers may provide them. Agricultural soil scientists investigate the movement of nutrients through the soil, and the amount of nutrients absorbed by a plant's roots. Agricultural soil scientists also examine the development of roots and their relation to the soil. Some agricultural soil scientists try to understand the structure and function of soils in relation to soil fertility. They grasp the structure of soil as porous solid. The solid frames of soil consist of mineral derived from the rocks and organic matter originated from the dead bodies of various organisms. The pore space of the soil is essential for the soil to become productive. Small pores serve as water reservoir supplying water to plants and other organisms in the soil during the rain-less period. The water in the small pores of soils is not pure water; they call it soil solution. In soil solution, various plant nutrients derived from minerals and organic matters in the soil are there. This is measured through the cation exchange capacity. Large pores serve as water drainage pipe to allow the excessive water pass through the soil, during the heavy rains. They also serve as air tank to supply oxygen to plant roots and other living beings in the soil. In short, agricultural soil scientists see the soil as a vessel, the most precious one for us, containing all of the substances needed by the plants and other living beings on earth.

Soil Preservation

In addition, agricultural soil scientists develop methods to preserve the agricultural productivity of soil and to decrease the effects on productivity of erosion by wind and water. For example, a technique called contour plowing may be used to prevent soil erosion and conserve rainfall. Researchers in agricultural soil science also seek ways to use the soil more effectively in addressing associated challenges. Such challenges include the beneficial reuse of human and animal wastes using agricultural crops; agricultural soil management aspects of preventing water pollution and the build-up in agricultural soil of chemical pesticides.

Employment of Agricultural Soil Scientists

Most agricultural soil scientists are consultants, researchers, or teachers. Many work in the developed world as farm advisors, agricultural experiment stations, federal, state or local government agencies, industrial firms, or universities. Within the USA they may be trained through the USDA's Cooperative Extension Service offices, although other countries may use universities, research institutes or research agencies. Elsewhere, agricultural soil scientists may serve in international organizations such as the Agency for International Development and the Food and Agriculture Organization of the United Nations.

Quotations

[The key objective of the soil science discipline is that of] finding ways to meet growing human needs for food and fiber while maintaining environmental stability and conserving resources for future generations

—John W. Doran, 2002 SSSA President, 2002

Many people have the vague notion that soil science is merely a phase of agronomy and deals only with practical soil management for field crops. Whether we like it or not this is the image many have of us

—Charles E. Kellog, 1961

Agrophysics

Agrophysics is a branch of science bordering on agronomy and physics, whose objects of study are the agroecosystem - the biological objects, biotope and biocoenosis affected by human activity, studied and described using the methods of physical sciences. Using the achievements of the exact sciences to solve major problems in agriculture, agrophysics involves the study of materials and processes occurring in the production and processing of agricultural crops, with particular emphasis on the condition of the environment and the quality of farming materials and food production.

Agrophysics is closely related to biophysics, but is restricted to the biology of the plants, animals, soil and an atmosphere involved in agricultural activities and biodiversity. It is different from biophysics in having the necessity of taking into account the specific features of biotope and biocoenosis, which involves the knowledge of nutritional science and agroecology, agricultural technology, biotechnology, genetics etc.

The needs of agriculture, concerning the past experience study of the local complex soil and next plant-atmosphere systems, lay at the root of the emergence of a new branch – agrophysics – dealing this with experimental physics. The scope of the branch starting from soil science (physics) and originally limited to the study of relations within the soil environment, expanded over time onto influencing the properties of agricultural crops and produce as foods and raw postharvest materials, and onto the issues of quality, safety and labeling concerns, considered distinct from the field of nutrition for application in food science.

Research centres focused on the development of the agrophysical sciences include the Institute of Agrophysics, Polish Academy of Sciences in Lublin, and the Agrophysical Research Institute, Russian Academy of Sciences in St. Petersburg.

Agricultural Biotechnology

Agricultural biotechnology, also known as agritech, is an area of agricultural science involving the use of scientific tools and techniques, including genetic engineering, molecular markers, molecular diagnostics, vaccines, and tissue culture, to modify living organisms: plants, animals, and microorganisms. Crop Biotechnology is one aspect of Agricultural Biotechnology which has been greatly developed upon in recent times. Desired trait are exported from a particular species of Crop to an entirely different species. These Transgene crops possess desirable characteristics in terms of flavor, color of flowers, growth rate, size of harvested products and resistance to diseases and pests.

Food Engineering

Food engineering is a multidisciplinary field of applied physical sciences which combines science, microbiology, and engineering education for food and related industries. Food engineering includes, but is not limited to, the application of agricultural engineering, mechanical engineering and chemical engineering principles to food materials. Food engineers provide the technological knowledge transfer essential to the cost-effective production and commercialization of food products and services. Physics, chemistry, and mathematics are fundamental to understanding and engineering products and operations in the food industry.

Bread factory in Germany

Food engineering encompasses a wide range of activities. Food engineers are employed in food processing, food machinery, packaging, ingredient manufacturing, instrumentation, and control. Firms that design and build food processing plants, consulting firms, government agencies, pharmaceutical companies, and health-care firms also employ food engineers. Specific food engineering activities include:

- drug/food products;

- design and installation of food/biological/pharmaceutical production processes;

- design and operation of environmentally responsible waste treatment systems;

- marketing and technical support for manufacturing plants.

Topics in Food Engineering

In the development of food engineering, one of the many challenges is to employ modern tools, technology, and knowledge, such as computational materials science and nanotechnology, to develop new products and processes. Simultaneously, improving quality, safety, and security remain critical issues in food engineering study. New packaging materials and techniques are being developed to provide more protection to foods, and novel preservation technology is emerging. Additionally, process control and automation regularly appear among the top priorities identified in food engineering. Advanced monitoring and control systems are developed to facilitate automation and flexible food manufacturing. Furthermore, energy saving and minimization of environmental problems continue to be important food engineering issues, and significant progress is being made in waste management, efficient utilization of energy, and reduction of effluents and emissions in food production.

Typical topics include:food people

- Advances in classical unit operations in engineering applied to food manufacturing

- Progresses in the transport and storage of liquid and solid foods

- Developments in heating, chilling and freezing of foods

- Advanced mass transfer in foods

- New chemical and biochemical aspects of food engineering and the use of kinetic analysis

- New techniques in dehydration, thermal processing, non-thermal processing, extrusion, liquid food concentration, membrane processes and applications of membranes in food processing

- Shelf-life, electronic indicators in inventory management, and sustainable technologies in food processing

- Modern packaging, cleaning, and sanitation technologies.

- Development of sensors systems for quality and safety assessment

Bioresource Engineering

Bioresource engineering is similar to biological engineering, except that it is based on biological and/or agricultural feedstocks. Bioresource engineering is more general and encompasses a wider range of technologies and various elements such as biomass, biological waste treatment, bioenergy, biotransformations and bioresource systems analysis, and technologies associated with Thermochemical conversion technologies such as combustion, pyrolysis, gasification, catalysis, etc.

Bioresource engineering also contains biochemical conversion technologies such as aerobic methods, anaerobic digestion, microbial growth processes, enzymatic methods, and composting. Products include fibre, fuels, feedstocks, fertilisers, building materials, polymers and other industrial products, and management products e.g. modelling, systems analysis, decisions, and support systems.

The impact of urbanization and increasing demand for food, water and land presents bioresource engineers with the task of bridging the gap between the biological world and traditional engineering. Agricultural and bioresource engineers attempt to develop efficient and environmentally sensitive methods of producing food, fiber, timber, bio-based products and renewable energy sources for an ever-increasing world population. Some of the research in bioresource engineering include machine vision, vehicle modification, wastewater irrigation, irrigation water management, stormwater management, inside natural environment for animals and plants, sensors, non-point source pollution and animal manure management.

Accomplishments

A biosynthesis of silver nanoparticles (NPs) mediated by fungal proteins of Coriolus versicolor has been undertaken for the first time last year. Hydrogels have been used to separate As(V) from water. The importance of bioresource engineers are usually obscured, but it is they who are the scientists helping the environment. Most recently bioresource engineers were working on checking the thermal properties of compost from waste to use it as a renewable, low temperature heat source.

ATCC

Founded in 1925, the ATCC (American Type Culture Collection) is a nonprofit and research organization, whose mission focuses on the acquisition, production, and development of standard reference microorganisms, cell lines and other materials for research in life sciences. ATCC has collected a wide range of biological items for research. Their holdings include molecular genomics tools, microorganisms and bioproducts.

Waste Management

Waste management is all the activities and actions required to manage waste from its inception to its final disposal. This includes amongst other things, collection, transport, treatment and disposal of waste together with monitoring and regulation. It also encompasses the legal and regulatory framework that relates to waste management encompassing guidance on recycling etc.

The term normally relates to all kinds of waste, whether generated during the extraction of raw materials, the processing of raw materials into intermediate and final products, the consumption of final products, or other human activities, including municipal (residential, institutional, commercial), agricultural, and social (health care, household hazardous wastes, sewage sludge). Waste management is intended to reduce adverse effects of waste on health, the environment or aesthetics.

Waste management in Kathmandu, Nepal

Waste management in Stockholm, Sweden

Waste management practices are not uniform among countries (developed and developing nations); regions (urban and rural area), and sectors (residential and industrial).

Central Principles of Waste Management

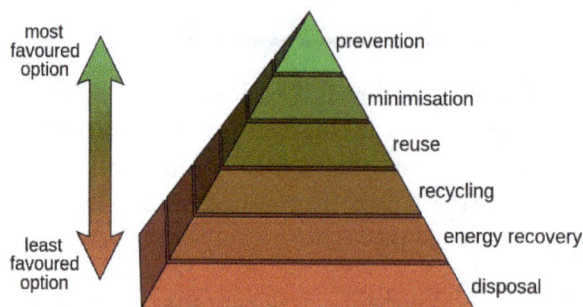
Diagram of the waste hierarchy

There are a number of concepts about waste management which vary in their usage between countries or regions. Some of the most general, widely used concepts include:

Waste Hierarchy

The waste hierarchy refers to the "3 Rs" reduce, reuse and recycle, which classify waste management strategies according to their desirability in terms of waste minimisation. The waste hierarchy remains the cornerstone of most waste minimisation strategies. The aim of the waste hierarchy is to extract the maximum practical benefits from products and to generate the minimum amount of waste; see: resource recovery. The waste hierarchy is represented as a pyramid because the basic premise is for policy to take action first and prevent the generation of waste. The next step or preferred action is to reduce the generation of waste i.e. by re-use. The next is recycling which would include composting. Following this step is material recovery and waste-to-energy. Energy can be recovered from processes i.e. landfill and combustion, at this level of the hierarchy. The final action is disposal, in landfills or through incineration without energy recovery. This last step is the final resort for waste which has not been prevented, diverted or recovered. The waste hierarchy represents the progression of a product or material through the sequential stages of the pyramid of waste management. The hierarchy represents the latter parts of the life-cycle for each product.

Life-cycle of a Product

The life-cycle begins with design, then proceeds through manufacture, distribution, use and then follows through the waste hierarchy's stages of reduce, reuse and recycle. Each of the above stages of the life-cycle offers opportunities for policy intervention, to rethink the need for the product, to redesign to minimize waste potential, to extend its use. The key behind the life-cycle of a product is to optimize the use of the world's limited resources by avoiding the unnecessary generation of waste.

Resource Efficiency

Resource efficiency reflects the understanding that current, global, economic growth and development can not be sustained with the current production and consumption patterns. Globally, we are extracting more resources to produce goods than the planet can replenish. Resource efficiency is the reduction of the environmental impact from the production and consumption of these goods, from final raw material extraction to last use and disposal. This process of resource efficiency can address sustainability.

Polluter Pays Principle

The Polluter pays principle is a principle where the polluting party pays for the impact caused to the environment. With respect to waste management, this generally refers to the requirement for a waste generator to pay for appropriate disposal of the unrecoverable material.

History

Throughout most of history, the amount of waste generated by humans was insignificant due to low population density and low societal levels of the exploitation of natural resources. Common waste produced during pre-modern times was mainly ashes and human biodegradable waste, and these were released back into the ground locally, with minimum environmental impact. Tools made out of wood or metal were generally re-used or passed down through the generations.

However, some civilizations do seem to have been more profligate in their waste output than others. In particular, the Maya of Central America had a fixed monthly ritual, in which the people of the village would gather together and burn their rubbish in large dumps.

Modern Era

Sir Edwin Chadwick's 1842 report The Sanitary Condition of the Labouring Population was influential in securing the passage of the first legislation aimed at waste clearance and disposal.

Following the onset of industrialisation and the sustained urban growth of large population centres in England, the buildup of waste in the cities caused a rapid deterioration in levels of sanitation and the general quality of urban life. The streets became choked with filth due to the lack of waste clearance regulations. Calls for the establishment of a municipal authority with waste removal powers occurred as early as 1751, when Corbyn Morris in London proposed that "... as the preservation of the health of the people is of great importance, it is proposed that the cleaning of this city, should be put under one uniform public management, and all the filth be...conveyed by the Thames to proper distance in the country".

However, it was not until the mid-19th century, spurred by increasingly devastating cholera outbreaks and the emergence of a public health debate that the first legislation on the issue emerged. Highly influential in this new focus was the report *The Sanitary Condition of the Labouring Population* in 1842 of the social reformer, Edwin Chadwick, in which he argued for the importance of adequate waste removal and management facilities to improve the health and wellbeing of the city's population.

In the UK, the Nuisance Removal and Disease Prevention Act of 1846 began what was to be a steadily evolving process of the provision of regulated waste management in London. The Metropolitan Board of Works was the first citywide authority that centralized sanitation regulation for the rapidly expanding city and the Public Health Act 1875 made it compulsory for every household to deposit their weekly waste in "moveable receptacles: for disposal—the first concept for a dust-bin.

Manlove, Alliott & Co. Ltd. 1894 destructor furnace. The use of incinerators for waste disposal became popular in the late 19th century.

The dramatic increase in waste for disposal led to the creation of the first incineration plants, or, as they were then called, "destructors". In 1874, the first incinerator was built in Nottingham by Manlove, Alliott & Co. Ltd. to the design of Albert Fryer. However, these were met with opposition on account of the large amounts of ash they produced and which wafted over the neighbouring areas.

Similar municipal systems of waste disposal sprung up at the turn of the 20th century in other large cities of Europe and North America. In 1895, New York City became the first U.S. city with public-sector garbage management.

Early garbage removal trucks were simply open bodied dump trucks pulled by a team of horses. They became motorized in the early part of the 20th century and the first close body trucks to eliminate odours with a dumping lever mechanism were introduced in the 1920s in Britain. These were soon equipped with 'hopper mechanisms' where the scooper was loaded at floor level and then hoisted mechanically to deposit the waste in the truck. The Garwood Load Packer was the first truck in 1938, to incorporate a hydraulic compactor.

Waste Handling and Transport

Molded plastic, wheeled waste bin in Berkshire, England

Waste collection methods vary widely among different countries and regions. Domestic waste collection services are often provided by local government authorities, or by private companies for industrial and commercial waste. Some areas, especially those in less developed countries, do not have formal waste-collection systems.

Waste Handling Practices

Curbside collection is the most common method of disposal in most European countries, Canada, New Zealand and many other parts of the developed world in which waste is collected at regular intervals by specialised trucks. This is often associated with curb-side waste segregation. In rural areas waste may need to be taken to a transfer station. Waste collected is then transported to an appropriate disposal facility. In some areas, vacuum collection is used in which waste is transported from the home or commercial premises by vacuum along small bore tubes. Systems are in use in Europe and North America.

Pyrolysis is used for disposal of some wastes including tires, a process that can produce recovered fuels, steel and heat. In some cases tires can provide the feedstock for cement manufacture. Such systems are used in USA, California, Australia, Greece, Mexico, the United Kingdom and in Israel. The RESEM pyrolysis plant that has been operational at Texas USA since December 2011, and processes up to 60 tons per day. In some jurisdictions unsegregated waste is collected at the curb-side or from waste transfer stations and then sorted into recyclables and unusable waste. Such systems are capable of sorting large volumes of solid waste, salvaging recyclables, and turning the rest into bio-gas and soil conditioner. In San Francisco, the local government established its Mandatory Recycling and Composting Ordinance in support of its goal of zero waste by 2020, requiring everyone in the city to keep recyclables and compostables out of the landfill.

The three streams are collected with the curbside "Fantastic 3" bin system – blue for recyclables, green for compostables, and black for landfill-bound materials – provided to residents and businesses and serviced by San Francisco's sole refuse hauler, Recology. The City's "Pay-As-You-Throw" system charges customers by the volume of landfill-bound materials, which provides a financial incentive to separate recyclables and compostables from other discards. The City's Department of the Environment's Zero Waste Program has led the City to achieve 80% diversion, the highest diversion rate in North America. Other businesses such as Waste Industries use a variety of colors to distinguish between trash and recycling cans.

Financial Models

In most developed countries, domestic waste disposal is funded from a national or local tax which may be related to income, or notional house value. Commercial and industrial waste disposal is typically charged for as a commercial service, often as an integrated charge which includes disposal costs. This practice may encourage disposal contractors to opt for the cheapest disposal option such as landfill rather than the environmentally best solution such as re-use and recycling. In some areas such as Taipei, the city government charges its households and industries for the volume of rubbish they produce. Waste will only be collected by the city council if waste is disposed in government issued rubbish bags. This policy has successfully reduced the amount of waste the city produces and increased the recycling rate.

Disposal Solutions

Landfill

A landfill compaction vehicle in action.

Spittelau incineration plant in Vienna

Incineration

Incineration is a disposal method in which solid organic wastes are subjected to combustion so as to convert them into residue and gaseous products. This method is useful for disposal of residue of both solid waste management and solid residue from waste water management. This process reduces the volumes of solid waste to 20 to 30 percent of the original volume. Incineration and other high temperature waste treatment systems are sometimes described as "thermal treatment". Incinerators convert waste materials into heat, gas, steam, and ash.

Incineration is carried out both on a small scale by individuals and on a large scale by industry. It is used to dispose of solid, liquid and gaseous waste. It is recognized as a practical method of disposing of certain hazardous waste materials (such as biological medical waste). Incineration is a controversial method of waste disposal, due to issues such as emission of gaseous pollutants.

Incineration is common in countries such as Japan where land is more scarce, as these facilities generally do not require as much area as landfills. Waste-to-energy (WtE) or energy-from-waste (EfW) are broad terms for facilities that burn waste in a furnace or boiler to generate heat, steam or electricity. Combustion in an incinerator is not always perfect and there have been concerns about pollutants in gaseous emissions from incinerator stacks. Particular concern has focused on some very persistent organic compounds such as dioxins, furans, and PAHs, which may be created and which may have serious environmental consequences.

Recycling

Recycling is a resource recovery practice that refers to the collection and reuse of waste materials such as empty beverage containers. The materials from which the items are made can be reprocessed into new products. Material for recycling may be collected separately from general waste using dedicated bins and collection vehicles, a procedure called kerbside collection. In some communities, the owner of the waste is required to separate the materials into different bins (e.g. for paper, plastics, metals) prior to its collection. In other communities, all recyclable materials are placed in a single bin for collection, and the sorting is handled later at a central facility. The latter method is known as "single-stream recycling."

Waste not the Waste. Sign in Tamil Nadu, India

The most common consumer products recycled include aluminium such as beverages cans, copper such as wire, steel from food and aerosol cans, old steel furnishings or equipment, rubber tyres, polyethylene and PET bottles, glass bottles and jars, paperboard cartons, newspapers, magazines and light paper, and corrugated fiberboard boxes.

Steel crushed and baled for recycling

PVC, LDPE, PP, and PS are also recyclable. These items are usually composed of a single type of material, making them relatively easy to recycle into new products. The recycling of complex products (such as computers and electronic equipment) is more difficult, due to the additional dismantling and separation required.

The type of material accepted for recycling varies by city and country. Each city and country has different recycling programs in place that can handle the various types of recyclable materials. However, certain variation in acceptance is reflected in the resale value of the material once it is reprocessed.

Re-use

Biological Reprocessing

An active compost heap.

Recoverable materials that are organic in nature, such as plant material, food scraps, and paper products, can be recovered through composting and digestion processes to decompose the organic matter. The resulting organic material is then recycled as mulch or compost for agricultural or landscaping purposes. In addition, waste gas from the process (such as methane) can be captured and used for generating electricity and heat (CHP/cogeneration) maximising efficiencies. The intention of biological processing in waste management is to control and accelerate the natural process of decomposition of organic matter.

Energy Recovery

Energy recovery from waste is the conversion of non-recyclable waste materials into usable heat, electricity, or fuel through a variety of processes, including combustion, gasification, pyrolyzation, anaerobic digestion, and landfill gas recovery. This process is often called waste-to-energy. Energy recovery from waste is part of the non-hazardous waste management hierarchy. Using energy recovery to convert non-recyclable waste materials into electricity and heat, generates a renewable energy source and can reduce carbon emissions by offsetting the need for energy from fossil sources as well as reduce methane generation from landfills. Globally, waste-to-energy accounts for 16% of waste management.

The energy content of waste products can be harnessed directly by using them as a direct combustion fuel, or indirectly by processing them into another type of fuel. Thermal treatment ranges from using waste as a fuel source for cooking or heating and the use of the gas fuel, to fuel for boilers to generate steam and electricity in a turbine. Pyrolysis and gasification are two related forms of thermal treatment where waste materials are heated to high temperatures with limited oxygen availability. The process usually occurs in a sealed vessel under high pressure. Pyrolysis of solid waste converts the material into solid, liquid and gas products. The liquid and gas can be burnt to produce energy or refined into other chemical products (chemical refinery). The solid residue (char) can be further refined into products such as activated carbon. Gasification and advanced Plasma arc gasification are used to convert organic materials directly into a synthetic gas (syngas) composed of carbon monoxide and hydrogen. The gas is then burnt to produce electricity and steam. An alternative to pyrolysis is high temperature and pressure supercritical water decomposition (hydrothermal monophasic oxidation).

Pyrolysis

Pyrolysis is a process of thermo-chemical decomposition of organic materials by heat in the absence of oxygen which produces various hydrocarbon gases. During pyrolysis, the molecules of object are subjected to very high temperatures leading to very high vibrations. Therefore, every molecule in the object is stretched and shaken to an extent that molecules starts breaking down. The rate of pyrolysis increases with temperature. In industrial applications, temperatures are above 430 °C (800 °F). Fast pyrolysis produces liquid fuel for feedstocks like wood. Slow pyrolysis produces gases and solid charcoal. Pyrolysis hold promise for conversion of waste biomass into useful liquid fuel. Pyrolysis of waste plastics can produce millions of litres of fuel. Solid products of this process contain metals, glass, sand and pyrolysis coke which cannot be converted to gas in the process.

Resource Recovery

Resource recovery is the systematic diversion of waste, which was intended for disposal, for a specific next use. It is the processing of recyclables to extract or recover materials and resources, or convert to energy. These activities are performed at a resource recovery facility. Resource recovery is not only environmentally important, but it is also cost effective. It decreases the amount of waste for disposal, saves space in landfills, and conserves natural resources.

Resource recovery (as opposed to waste management) uses LCA (life cycle analysis) attempts to offer alternatives to waste management. For mixed MSW (Municipal Solid Waste) a number of broad studies have indicated that administration, source separation and collection followed by reuse and recycling of the non-organic fraction and energy and compost/fertilizer production of the organic material via anaerobic digestion to be the favoured path.

As an example of how resource recycling can be beneficial, many of the items thrown away contain precious metals which can be recycled to create a profit, such as the components in circuit boards. Other industries can also benefit from resource recycling with the wood chippings in pallets and other packaging materials being passed onto sectors such as the horticultural profession. In this instance, workers can use the recycled chips to create paths, walkways, or arena surfaces.

Sustainability

The management of waste is a key component in a business' ability to maintaining ISO14001 accreditation. Companies are encouraged to improve their environmental efficiencies each year by eliminating waste through resource recovery practices, which are sustainability-related activities. One way to do this is by shifting away from waste management to resource recovery practices like recycling materials such as glass, food scraps, paper and cardboard, plastic bottles and metal. This topic was on the agenda of the international Conference on Green Urbanism, held in Italy 12–14 October 2016.

Avoidance and Reduction Methods

An important method of waste management is the prevention of waste material being created, also known as waste reduction. Methods of avoidance include reuse of second-hand products, repairing broken items instead of buying new, designing products to be refillable or reusable (such as cotton instead of plastic shopping bags), encouraging consumers to avoid using disposable products (such as disposable cutlery), removing any food/liquid remains from cans and packaging, and designing products that use less material to achieve the same purpose (for example, lightweighting of beverage cans).

International Waste Movement

While waste transport within a given country falls under national regulations, trans-boundary movement of waste is often subject to international treaties. A major concern to many countries in the world has been hazardous waste. The Basel Convention, ratified by 172 countries, deprecates movement of hazardous waste from developed to less developed countries. The provisions of the Basel convention have been integrated into the EU waste shipment regulation. Nuclear waste, although considered hazardous, does not fall under the jurisdiction of the Basel Convention.

Benefits

Waste is not something that should be discarded or disposed of with no regard for future use. It can be a valuable resource if addressed correctly, through policy and practice. With rational and consistent waste management practices there is an opportunity to reap a range of benefits. Those benefits include:

1.   Economic – Improving economic efficiency through the means of resource use, treatment and disposal and creating markets for recycles can lead to efficient

practices in the production and consumption of products and materials resulting in valuable materials being recovered for reuse and the potential for new jobs and new business opportunities.

2. Social – By reducing adverse impacts on health by proper waste management practices, the resulting consequences are more appealing settlements. Better social advantages can lead to new sources of employment and potentially lifting communities out of poverty especially in some of the developing poorer countries and cities.

3. Environmental – Reducing or eliminating adverse impacts on the environmental through reducing, reusing and recycling, and minimizing resource extraction can provide improved air and water quality and help in the reduction of greenhouse gas emissions.

4. Inter-generational Equity – Following effective waste management practices can provide subsequent generations a more robust economy, a fairer and more inclusive society and a cleaner environment.

Challenges in Developing Countries

Waste management in cities with developing economies and economies in transition experience exhausted waste collection services, inadequately managed and uncontrolled dumpsites and the problems are worsening. Problems with governance also complicate the situation. Waste management, in these countries and cities, is an ongoing challenge and many struggle due to weak institutions, chronic under-resourcing and rapid urbanization. All of these challenges along with the lack of understanding of different factors that contribute to the hierarchy of waste management, affect the treatment of waste.

Technologies

Traditionally the waste management industry has been a late adopter of new technologies such as RFID (Radio Frequency Identification) tags, GPS and integrated software packages which enable better quality data to be collected without the use of estimation or manual data entry.

References

- Lovelock, James (2009). The Vanishing Face of Gaia: A Final Warning: Enjoy It While You Can. Allen Lane. ISBN 978-1-84614-185-0.

- Encyclopedia of Agrophysics in series: Encyclopedia of Earth Sciences Series edts. Jan Glinski, Jozef Horabik, Jerzy Lipiec, 2011, Publisher: Springer, ISBN 978-90-481-3585-1

- Encyclopedia of Soil Science, edts. Ward Chesworth, 2008, Uniw. of Guelph Canada, Publ. Springer, ISBN 978-1-4020-3994-2

- Physical Methods in Agriculture. Approach to Precision and Quality, edts. J. Blahovec and M. Kutilek, Kluwer Academic Publishers, New York 2002, ISBN 0-306-47430-1.

- Soil Physical Condition and Plant Roots by J. Gliński, J. Lipiec, 1990, CRC Press, Inc., Boca Raton, USA, ISBN 0-8493-6498-1

- Soil Aeration and its Role for Plants by J. Gliński, W. Stępniewski, 1985, Publisher: CRC Press, Inc., Boca Raton, USA, ISBN 0-8493-5250-9

- Scientific Dictionary of Agrophysics: polish-English, polsko-angielski by R. Dębicki, J. Gliński, J. Horabik, R. T. Walczak - Lublin 2004, ISBN 83-87385-88-3.

- Education Portal (2014). "Careers in Agricultural Economics: Job Options and Requirements". Retrieved 2014-10-11.

- Anthony P. Carnevale; Jeff Strohl; Michelle Melton (2011). "What's It Worth? The Economic Value of College Majors". Retrieved 2014-10-11.

- Culas, Richard and Mahendrarajah Causes of Diversification in Agriculture over Time: Evidence from Norwegian Farming Sector, 2005. (Retrieved on 2011-9-27).

- Singh , R Paul; Dennis R. Heldman (2013). Introduction to Food Engineering (5th ed.). Academic Press. p. 1. ISBN 0123985307.

- Guidelines for National Waste Management Strategies Moving from Challenges to Opportunities (PDF). United Nations Environmental Programme. 2013. ISBN 978-92-807-3333-4..

- Claire Swedberg (4 February 2014). "Air-Trak Brings Visibility to Waste Management". RFID Journal. Retrieved 1 October 2015.

- Grossi, M.; Di Lecce, G.; Gallina Toschi, T.; Riccò, B. (2014). "A novel electrochemical method for olive oil acidity determination". Microelectronics Journal. 45: 1701–1707.

- "Editorial Board/Aims & Scope". Waste Management. 34 (3): IFC. March 2014. doi:10.1016/S0956-053X(14)00026-9.

- Barbalace, Roberta Crowell (2003-08-01). "The History of Waste". EnvironmentalChemistry.com. Retrieved 2013-12-09.

- National Waste & Recycling Association. "History of Solid Waste Management". Washington, D.C. Retrieved 2013-12-09.

Permissions

All chapters in this book are published with permission under the Creative Commons Attribution Share Alike License or equivalent. Every chapter published in this book has been scrutinized by our experts. Their significance has been extensively debated. The topics covered herein carry significant information for a comprehensive understanding. They may even be implemented as practical applications or may be referred to as a beginning point for further studies.

We would like to thank the editorial team for lending their expertise to make the book truly unique. They have played a crucial role in the development of this book. Without their invaluable contributions this book wouldn't have been possible. They have made vital efforts to compile up to date information on the varied aspects of this subject to make this book a valuable addition to the collection of many professionals and students.

This book was conceptualized with the vision of imparting up-to-date and integrated information in this field. To ensure the same, a matchless editorial board was set up. Every individual on the board went through rigorous rounds of assessment to prove their worth. After which they invested a large part of their time researching and compiling the most relevant data for our readers.

The editorial board has been involved in producing this book since its inception. They have spent rigorous hours researching and exploring the diverse topics which have resulted in the successful publishing of this book. They have passed on their knowledge of decades through this book. To expedite this challenging task, the publisher supported the team at every step. A small team of assistant editors was also appointed to further simplify the editing procedure and attain best results for the readers.

Apart from the editorial board, the designing team has also invested a significant amount of their time in understanding the subject and creating the most relevant covers. They scrutinized every image to scout for the most suitable representation of the subject and create an appropriate cover for the book.

The publishing team has been an ardent support to the editorial, designing and production team. Their endless efforts to recruit the best for this project, has resulted in the accomplishment of this book. They are a veteran in the field of academics and their pool of knowledge is as vast as their experience in printing. Their expertise and guidance has proved useful at every step. Their uncompromising quality standards have made this book an exceptional effort. Their encouragement from time to time has been an inspiration for everyone.

The publisher and the editorial board hope that this book will prove to be a valuable piece of knowledge for students, practitioners and scholars across the globe.

Index

www.ingramcontent.com/pod-product-compliance
Lightning Source LLC
Chambersburg PA
CBHW061932190326
41458CB00009B/2718

* 9 7 8 1 6 3 5 4 9 0 1 7 6 *